Sustainable Bamboo Development

Sustainable Bamboo Development

Zhu Zhaohua

*Former Professor and (retired) Senior Researcher, Chinese Academy of Forestry
Distinguished Fellow of the International Bamboo and Rattan Organization (INBAR)
Honourable Chairman of the Academic Association of Forest Ecology, Chinese
Society of Forestry*

and

Jin Wei

*Training Coordinator of the International Bamboo and Rattan Organization
(INBAR, originally the International Network for Bamboo and Rattan)*

CABI is a trading name of CAB International

CABI	CABI
Nosworthy Way	745 Atlantic Avenue
Wallingford	8th Floor
Oxfordshire OX10 8DE	Boston, MA 02111
UK	USA
Tel: +44 (0)1491 832111	Tel: +1 (617)682-9015
Fax: +44 (0)1491 833508	E-mail: cabi-nao@cabi.org
E-mail: info@cabi.org	
Website: www.cabi.org	

The designations employed and the presentation of the material in this publication do not imply the expression of any opinion whatsoever on the part of CABI concerning the legal status of any country, territory, city of area or of its authorities, or concerning the delimitation of its frontiers or boundaries. Lines on maps represent approximate border lines for which there may not yet be full agreement.

A catalogue record for this book is available from the British Library, London, UK.

Library of Congress Cataloging-in-Publication Data

Names: Zhu, Zhaohua, author. | Jin, Wei, author.
Title: Sustainable bamboo development / Zhu Zhaohua, Jin Wei.
Description: Boston, MA : CABI, [2018] | Includes bibliographical references and index.
Identifiers: LCCN 2017029485 (print) | LCCN 2017031464 (ebook) | ISBN 9781786394026 (pdf) | ISBN 9781786394033 (ePub) | ISBN 9781786394019 (hardback : alk. paper)
Subjects: LCSH: Bamboo. | Bamboo--Economic aspects.
Classification: LCC SB317.B2 (ebook) | LCC SB317.B2 Z485 2018 (print) | DDC 633.5/8--dc23
LC record available at https://lccn.loc.gov/2017029485

ISBN-13: 9781786394019

Commissioning editor: Rachael Russell
Editorial assistant: Alexandra Lainsbury
Production editor: James Bishop

Typeset by SPi, Pondicherry, India
Printed and bound by CPI Group (UK) Ltd, Croydon, CRO 4YY

Contents

_____

Brief Introduction to the Authors

Zhu Zhaohua

Prof. Zhu Zhaohua graduated from the Faculty of Botanical Ecology, Lanzhou University, China in 1962. He is a former Deputy Director of the Forestry Research Institute of the Chinese Academy of Forestry (CAF), and a former Director of the International Farm Forestry Training Institute of CAF. He is also a former Director of the Department of International Cooperation of the CAF, Honourable Chairman of the Academic Association of Forest Ecology, Chinese Society of Forestry, former Deputy Director General of INBAR (the International Network for Bamboo and Rattan, known since May 2017 as the International Bamboo and Rattan Organization) and an INBAR Distinguished Fellow.

As a Senior Researcher who has conducted forestry science over more than 50 years, Prof. Zhu Zhaohua has especially important achievements in research on the rehabilitation of natural secondary forests, the rehabilitation of degraded lands, the optimization of agroforestry models, *Paulownia* cultivation technology and genetic improvement, integrated mountain sustainable development, and key issues affecting the sustainable developments of non-timber forestry products and bamboo sectors. He has had many national and international awards, of which some of the most outstanding are: designation by the China State Council in 1991 as 'The Special Contribution Scientist of State'; receipt of the first award of the title of 'Man of the Trees' by the International Richard St. Barbe Baker Foundation (formerly 'the Men of the Trees', now the International Tree Foundation) for contributions to developing rural forestry in China (1985); designation by the then Prime Minister of Canada – Jean Chrétien – on the occasion of the 25th Anniversary of the International Development Research Centre (IDRC) as a world outstanding scientist who has contributed to leading international cooperation projects (1995); the titles of All Life Research Professor of Indian Academy of Forestry (1995) and Honorable Member of the American Society of Foresters (2002); and the bestowing of the 'World Bamboo Pioneer Award' by World Bamboo Organization at the 10th World Bamboo Congress in South Korea (2015).

From 1987 to 1998, Prof. Zhu Zhaohua organized eight international training workshops on agroforestry and multipurpose tree species (1990–1997) and organized 4 training courses on Forest Ecology for Cottbus University of Germany. From 1999 to 2015, with the sponsorship of The China Ministry of Science and Technology (MOST), he conducted 19 training workshops on bamboo and rattan development, and at the request of various international organizations and countries, he provided 25 training courses and study tours on bamboo and rattan development. The total number of participants in the above training activities amounted to over 1400 professionals from 64 countries. During the past 38 years, he has also been invited to visit 26 countries worldwide to provide consultancy services on agroforestry, bamboo and rattan development.

Prof. Zhu Zhaohua's e-mail address is zhzhubamb@163.com

Professor Zhu Zhaohua

Jin Wei

Jin Wei has an MA in Journalism from the School of Journalism & Communications, Peking University. She has 19 years of experience working in international development and capacity building for developing countries in INBAR, and has accumulated a certain amount of knowledge and experience in non-timber forest product development and rural sustainable development, as well as in related policy making and strategy identification. She has taken on major responsibilities in the organization of more than 30 international seminars and training workshops, in which total number of international participants exceeded 1000. Jin Wei also has16 years of experience in organizational public relations development, and much experience in conferences and publications.

Jin Wei's e-mail address is wjin@inbar.int

Jin Wei

Other Contributors

_____

Ramesh Chandra Chaturvedi, Saket, 11 Prem Nagar, Sapru Marg, Lucknow 226001, India. E-mail: cramesh1937@yahoo.co.in

Ximena Londoño, Carrera 5 #14-26, Montenegro, Quindío, Colombia. E-mail: ximelon@gmail.com

Romualdo L. Sta Ana, 7-2425 Portage Avenue, Winnipeg R3J 0N2, Canada. E-mail: romisantana@yahoo.com

Forewords

_____

Foreword I

A healthy and stable environment is vital for survival of humanity, but this is under threat from growing levels of greenhouse gas emissions, land degradation, desertification, deforestation, biodiversity loss, pollution, poverty and other forms of degradation. Sustainability is imperative, not a choice. Here, sustainable bamboo development through an integrated approach to economic and social development and environmental protection seems to be a viable path forward in which living conditions and resource use continue to meet human needs without undermining the 'integrity, stability and beauty' of the natural biotic system.

INBAR (the International Network for Bamboo and Rattan, known since May 2017 as the International Bamboo and Rattan Organization) was created to realize the potential of bamboo and rattan for global sustainability, and as a means of achieving a pro-poor green economy and the development of a network based on South–South Cooperation. Other international organizations, such as the World Bamboo Congress (WBC) and the World Bamboo Organization (WBO), have also been playing important international roles for establishing dialogue and exchange platforms for bamboo development and research, promoting the popularization of bamboo knowledge and technologies, bamboo innovations and pro-poor trading systems.

I am pleased to provide a quick preface to the book _Sustainable Bamboo Development_ by renowned author Prof. Zhu Zhaohua, who has worked with the Chinese Academy of Forestry (CAF) and INBAR. His contributions to sustainable development and poverty alleviation in China have been recognized by many national and international accolades and awards. He has been especially recognized internationally for his excellent work in developing a scientific agroforestry system in the North China Plain, initiating non-timber forest product development in the mountainous areas of Lin'an County in Zhejiang Province as one of China's earliest models of rural mountainous sustainable development, and establishing the bamboo and rattan industry in Hainan Province.

Prof. Zhu Zhaohua was a pioneer in the development of China's bamboo sector and facilitated its progressive development from simple raw material production to extended industrial processing, engineering utilization, to high added-value, high-quality and environmentally friendly products. Towards achieving the goals of sustainable development, the processing industry, ecotourism and the cultural industries are mutually supporting each other in reducing poverty, hunger and inequality, combating climate change and improving sustainability.

Endemic bamboo species are found worldwide on all continents except for Europe and Antarctica. The Asia–Pacific region has the largest area of bamboo and the resource is important in socio-economic, sociocultural, socioecological, climatic and functional contexts. As an enduring, versatile and renewable resource, bamboo can contribute significantly to sustainable development, poverty eradication and countering climate change.

The bamboo-based industry is experiencing a real surge. The major recent uses of bamboo for construction, furniture, flooring, paper, textiles, medicines, charcoal, food and beverages etc., illustrate the versatility and potential of this plant. China has taken a lead in research and innovation, production, processing and trade related to bamboo. It has already trained 1180 people from 60 countries.

Many developing countries across the world are looking at the bamboo sector as a solution to several sustainable development goals (SDGs) – poverty eradication (SDG 1), access to safe water (SDG 6.1) decent work and economic growth (SDG 8), industrial innovation (SDG 9), climate actions (SDG 13) and sustainability (SDG 15). This book will leverage existing strengths and knowledge, combined with technological, policy and institutional innovations, to turn bamboo into a material for a sustainable future as a part of a pro-poor, green economy.

The book contains excellent reference material for practitioners and scientists alike, and I am confident that it will promote sustainable bamboo development, the reduction of poverty and protection of the environment, as well as enabling people to live a life of dignity.

Dr Dina Nath Tewari
President
Utthan Centre for Sustainable Development and Poverty Alleviation, India

Foreword II

When readers venture into this book, they will find very varied and specialized contents on bamboo – from its world distribution to case studies by continent and by country in the bamboo sector. When they have had the chance to read and study some of the chapters in detail, they will realize the value of this compilation by Prof. Zhu Zhaohua – a compilation that is expected to provide a basis for politicians, planners, scientists from different areas and the public in general to be able to achieve a sustainable industrialization process for bamboo, when they have become familiar with the various components, processes and ways in which the bamboo chain operates in different countries of the world.

As a specialized reader, there were two aspects that strongly attracted my attention. The first is that the author is a person with vast experience acquired through the consulting work he has done and the specialized guidance he has provided to more than 22 bamboo-growing countries in Asia, Latin America and Africa, a person who has also received provincial, national and international awards and recognitions. Among these stand out the 'Man of the Trees' award granted by the (then) International Richard St. Barbe Baker Foundation for his extraordinary contributions to the development of a specialized agroforestry system in China that consisted of large areas of crops intercropped with *Paulownia* in the North Plain Agriculture Area of China. Because Prof. Zhu Zhaohua was the first 'Man of the Trees', a ceremony granting him this honour was held at the headquarters of the United Nations in 1985. He was also granted the 'Bamboo Pioneer 2015' award during the 10th World Bamboo Congress in South Korea for his leadership in that sector. The second aspect of the book is the integrated vision that the author has on bamboo, and his personal conviction – and experience – of being able to achieve the sustainable industrial development of bamboo, which he conveys through the analysis of both successful and unsuccessful cases in several parts of the world.

In these historic times, the sustainable development of bamboo is a common-sense premise, although the way to achieve this is an ongoing debate in academic, technical and political media. In Latin America, this clash of opinions has been dominated by lack of both information and the relevant scientific knowledge translated into terms that allow political debate on and implementation of sustainable bamboo development within the economic context of the country concerned. Translating these scientific concepts into policies and legislation is one of the most difficult issues now faced by Latin American countries that are rich in bamboo diversity but poor in the social and economic development of their rural sector.

The testimonies captured in this book on the sustainable industrial development of bamboo and the collateral benefits for the environment, and for the social, economic and cultural aspects of this, will help to demonstrate to the people – and to the institutions that rule politics, legislation and the enforcement of the law in our Latin countries – that communities can be transformed through the cultivation of bamboo and that bamboo cultivation can help to close the gap that separates development from the adequate use of resources and their conservation.

I invite you to read and learn from masterly work.

Ximena Londoño
President of Colombia Bamboo Society

Foreword III

As argued widely, bamboo is one of the most versatile forest species; it also has a very high potential for value addition. Bamboo development is at different phases in the various regions of the world where it grows naturally. Africa in particular lags behind the Asian and Latin American continents in the development of bamboo as a resource. Interest in bamboo development in Africa has existed for several decades, but the amount of effort that has been put into this has been isolated, low and slow. For example, in the 1950s and 1960s, several countries established exotic bamboo species in various forms, including arboretums (Ghana, Kenya, Malawi, Rwanda, Tanzania), street plantings in towns (Ghana, Kenya, Nigeria, South Africa), planting around Orthodox Church compounds (notably in Ethiopia) and general bamboo groves planted in forest premises (several countries). In all of these cases, what has notably lacked is a focus on proper management for the sustainable production of culms for different products and uses.

It is, however, notable that in the 1980s research and extension support for bamboo development started in Africa. Training in bamboo utilization was undertaken mainly in Ethiopia, with some nursery and growth production research also recorded in Kenya and Ghana. Funding support was extended into the 1990s mainly by the International Development Research Centre (IDRC) of Canada and China, and their close partners. The establishment of the INBAR (the International Network for Bamboo and Rattan, known since May 2017 as the International Bamboo and Rattan Organization) enhanced bamboo activities in Africa, especially from 2000, mainly with a focus on production, and to a lesser extent on utilization, mostly on exotic species introduced into Africa. It was only later in the 2000s that research and development of the main and widely distributed indigenous African bamboo species *Oxytenanthera abyssinica* and *Yushania alpina* (syn. *Arundinaria alpina*) was embarked upon. In addition to this, African technical officers benefited greatly from the capacity improvement for bamboo production and utilization driven by the Chinese government, INBAR, the European Union (EU) and other international organizations during this period.

It is apparent from the above that the awareness and capacity for bamboo value chain development is still lacking in Africa if it is to fully benefit from the great potential of bamboo in livelihood enhancement and wealth creation. Prof. Zhu Zhaohua's book on *Sustainable Bamboo Development* is a perfect fit for global bamboo developers and investors, but more so for Africa, which lags behind other tropical bamboo growers and users. As a priority in Africa, there is need to create more awareness of bamboo and its development potential. Those new to bamboo practices and the bamboo industry will need to understand the biology and inherent characteristics of bamboo and where it fits as a commodity into human development and environmental management; this is an aspect that Prof. Zhu Zhaohua covers well in the first chapter of his book.

China, as the global giant of bamboo development, provides the best statistics of case studies for benchmarking bamboo development. The book takes deliberate advantage of this fact and guides the reader with viable examples through a value chain development scenario. Prof. Zhu Zhaohua elaborates on the requirements of a growing bamboo industry, and points out the importance of understanding the bamboo resource base, planning, the prioritization of product development according to the demands of the country concerned, enhancing technologies, putting the right policies in place to drive the national bamboo industry and concomitant capacity building – including 'multi-participation' by the various stakeholders involved in the bamboo industry in a country. Also demonstrated, with enhanced case studies, is the need for a market development and monitoring and evaluation model that includes improvement and innovation opportunities for a growing national bamboo industry. The role of the government in shaping bamboo sector development is emphasized, with a caution to governments not to take over management of the actual enterprises.

Prof. Zhu Zhaohua has long experience in research and development, management and capacity building for bamboo, and this is ably demonstrated by the contents of the book. The volume is reader friendly, and the use of visual aids in the form of elaborate and clear photos, global case studies of best practices and benchmarked operational statistics from the industry add value to the intended level of communication. The book vividly discusses the technical, economic and policy issues surrounding the growing of bamboo and related enterprises. It also touches upon many of the aspects of bamboos in human development and the author's association with bamboo, while at the same time cautioning on the practical pitfalls that can lead to failures in an intended developing bamboo industry.

Due to its wide coverage in bamboo development aspects, the book is of benefit and is therefore recommended for reading by bamboo field technicians, researchers, farmers, potential and actual bamboo investors, development partners and national and international policy makers.

Bernard N. Kigomo (PhD)
Senior Deputy Director, Kenya Forestry Research Institute

Preface

Before the 1980s, the bamboo growing and bamboo industrial processing had not yet caught universal attention. The mechanical processing of bamboo poles had only been developed in Japan and China's Taiwan area and the bamboo pulping industries had been developed in China, India and Brazil, but the majority of bamboo-producing countries were utilizing bamboos in traditional, manual ways. After the 1980s, under the leadership and support of the International Development Research Centre (IDRC) of Canada, a number of bamboo-producing countries started some research and development work on bamboo, and various bamboo-related international organizations, such the World Bamboo Organization (WBO), the World Bamboo Congress (WBC) and INBAR (the International Network for Bamboo and Rattan, known since May 2017 as the International Bamboo and Rattan Organization) were established. Bamboos were then being promoted and put into the view of international society.

Since 1987, the IDRC, in collaboration with the China Ministry of Science and Technology (MOST), has started to sponsor international bamboo training workshops. The author worked with the Chinese Academy of Forestry (CAF), the Lin'an Modern Forestry Science and Technology Center (LMFSTC) in Zhejiang Province and INBAR in undertaking these workshops. Over the past 19 years, 19 workshops and 25 study tours have been carried out, based on the needs and requirements of bamboo-producing countries and international organizations. Altogether, 1180 participants from more than 60 countries were trained under this programme. In recent years, owing to increasing demand, more training workshops have been organized by the China National Bamboo Research Center (CBRC) and the China International Bamboo and Rattan Center (ICBR). These series of capacity building and training activities have largely raised people's awareness on the significance of bamboos and of bamboo industry development. At the same time, the activities have shared the successful experiences of the development of China's bamboo sector.

However, there have been serious gaps among the bamboo sectors of different countries. In a large number of bamboo-producing countries, the bamboo forests were still untouched natural resources, which could hardly be utilized immediately for industrial purpose. Policies in favour of bamboo development were lacking in most bamboo-producing countries, and sometimes certain existing country policies even had negative effects. Also, due to lack of experience, many investments in bamboo development failed.

The purpose of the book is to analyse and summarize the experiences of successful and failed cases in China and other countries in Asia, Africa and Latin America and, based on the analysis and summaries, provide some key views and suggestions on how to develop the bamboo sector in a sustainable way. All the experiences and practices that are described are limited in scope, the basic principle here being to identify practical plans according to real conditions, and to strategically study the experience of and lessons learned in different countries.

The second chapter of the book introduces bamboos as plants that have very special features, and make contributions to ecology, the environment and human beings that are far more than people can imagine. In Asia, especially in East Asia, bamboo has been a symbol of the East Asian culture. In Latin America, there is a strong culture in the utilization of Guadua bamboo (*Guadua angustifolia*).

A brief analysis of – and introduction to – the features of bamboo plants is given, including their strong renewability, their fast-growing and high biomass production characteristics, and their important ecological functions, including water and soil conservation, role as a carbon sink and in carbon storage, and adaptability to climate change. In China, the bamboo forest area is only 2.78% of the total forest area, but the carbon storage of bamboo forests is 11.6% of the total. Bamboos, as plant resources, are very easy to manage, and have beautiful landscaping and ornamental effects. Based on research on their physical and chemical features, bamboos could widely replace other timber materials and, in certain cases, they are advantageous. They have great potential in terms of their economic value.

In Anji County of Zhejiang Province, China, more than 3000 types of bamboo products in nine series were being produced, including structural and decorative materials made from bamboo-based panels, bamboo curtains, mats and carpets, fibre products, biochemical products from bamboo leaves and culms, such as medicines, beverages (such as tea) etc., bamboo charcoal and other biomass energy products, bamboo machinery, bamboo shoots and other bamboo-related food products, bamboo handicrafts, etc. In 2015, the total production value of Anji's bamboo processing industry reached 13.78 billion CNY (Chinese yuan), and the bamboo forest-based ecotourism and cultural industry created a value of 5.22 billion CNY. The number of tourists to Anji reached 34 times the number of local residents – up to 11 million. The total production value of the bamboo sector reached 19 billion CNY (3.065 billion US$), which is 12% of China's total bamboo production value. Bamboo forest-based ecotourism and the bamboo cultural industry have become a new growing point of China's bamboo sector, and the multifunctional characteristics and effects of bamboo forests are now being gradually explored. The development of China's bamboo sector has experienced a transformational period from simple raw production to extended industrial processing, engineering utilization, and then to high-added value, high-quality, environmentally friendly products; it is now in a new stage where the multiple functions and effects of the bamboo forests are being fully explored, and the processing industry, ecotourism and the bamboo cultural industry are mutually supporting each other, forming a new harmonized development stage.

The book also gives a brief introduction to the rich connotation of Chinese bamboo culture. In Chinese culture, bamboos can not only supply all kinds of living necessities, but are also recognized as a symbol of fine virtue and morality. China's bamboo culture has been continuously inherited and carried forward in the country's 5000 year history. It has rich connotations and has had great impacts on China's writing characters, literature, poems, music, dance, paintings and handicrafts. Bamboos were widely used and made great contributions to China's history in people's daily lives, in construction, textiles, the pulp and papermaking industry, transportation, agriculture and irrigation, weapons and war facilities, mining, furniture making, food and medicines, gardening and landscaping, etc. To sum up, bamboos had deep impacts on China's, East Asia's and Latin America's spiritual civilization and material civilization. In this part of the book, we also invited Prof. Ramesh Chandra Chaturvedi, a Senior Professor of bamboo engineering and utilization, to present a historical section that introduces bamboo utilization and development history, and its impacts on the local people's culture in India.

The third chapter of the book discusses the bamboo sector. In comparison with other processing industries, the bamboo sector is much more complicated. It consists of a comprehensive supply chain from resource cultivation and management, to industrial processing and market sales. It is necessary that all of the stakeholders – bamboo farmers, processing enterprises, scientists and technicians, policy makers – actively participate and closely cooperate. While developing the bamboo sector, attention needs to be paid to protection of the ecological system and community economy development in order to achieve a multiple win situation. Although the application technologies are not very complicated, maintenance of the sustainability of the whole sector is not an easy task.

How to develop the bamboo sector in a sustainable way? Based on my experience gathered through study tours and consultancies in 22 bamboo-producing countries in Asia, Latin America and Africa, and organizing a series of international training workshops and study tours since 1999, and in combination with my many years of research and analysis on quite a number of successful and failed cases in China and other countries, I (herein after refer to the first Author – Prof. Zhu Zhaohua) am able to describe eight key issues that have an impact on the sustainability of a bamboo sector.

1. Bamboo resources are the basis of a sustainable bamboo sector. Before industrial processing, it is a must to learn about the production, area, distribution, species structure and provenances of the local bamboo resources. Moreover, it is also necessary to identify the features of the local bamboo species so as to know what type of products they are suitable for, as well as to collect other necessary data for preparing for their industrial utilization; this information includes the quality and quantity of resources that can be immediately accessed and utilized for industrial purposes. Many failures have been caused by a lack of thorough studies and evaluations of the local bamboo resources. In the book, we have introduced the case of a bamboo flooring factory in Ecuador, which failed to sustain itself due to lack of a thorough study on the locally available bamboo resources.

2. A proper and practical development plan is the guarantee for success and for avoiding any detour. No matter whether it is a country, a region, a village or an enterprise, before implementing a bamboo project, it is very necessary to make a good plan. The planning should include all aspects of the bamboo sector: (i) resource development and protection, which includes the scale and structure of the bamboo resources; (ii) short-, medium- and long-term development goals and their necessary conditions; (iii) relative government policies; (iv) land, technologies and facilities to be in place; (v) personnel training; (vi) estimates of investment and outputs; and (vii) the construction of demonstration sites and markets, etc. The plan should be identified by means of the multiple participation (multi-participation) of all stakeholders. A good plan may count for 50% of future success. Blind investment or rushed investment without deliberation often results in failure and detours. New plans should be identified along with the development course of the bamboo sector, so as to guide the sector to develop to higher stages. This book has used the 5 year bamboo development plan of Lin'an County, Zhejiang Province, and China's National Bamboo Development Plan, 2013–2020, as examples.

3. Product development and optimization of the product structure. Based on the status of local bamboo resources and traditional products, and according to market demands, local traditional products should be the first to be identified and developed. This will allow the advantages of the local species to be fully explored, and at the same time, it will allow markets to be opened and accessed. While developing local products, the biomass production and the improvement of biomass utilization rate should be identified as major issues to be tackled. Monocultural production will inevitably cause a low utilization rate of raw materials. In this type of production, the bamboo utilization rate is usually lower than 20%, and a large amount of bamboo materials are wasted. At present, in China's Anji County, the processing industry can fully utilize the whole bamboo culm, the shoot, the sheath, the leaves, the rhizomes and the roots, as well as the residues from engineering processing, such as sawdust and slices, etc. As a result, the bamboo biomass utilization rate in Anji County is 85–90%, compared with 20% in the past. Thousands of different products are made out of bamboo in Anji County, and the value chain has been extending continuously. The optimization of bamboo product structure to meet market needs is also very necessary. Continuous innovation is key for the development of higher value products. It is especially important to note that since 2000, China has started to pay attention to the development of ecological and cultural products from bamboo forests, and staying on farms and ecotourism in the bamboo forests have been developing fast, which has escalated bamboo product structure and the development of the bamboo sector. The book describes the structural optimization of bamboo products in Anji County at different development stages of the bamboo sector.

4. The construction of demonstration sites is a guarantee of success. Practice has proved that the power of demonstration modes is immeasurable. In the course of bamboo sector development, it is usually necessary to implement trials within a certain scale or at certain demonstration sites. When the trials are successful, the experience gained can be extended to other sites. While constructing a demonstration base, one should note the following:

- Multi-participation is a necessity. The participation of enterprises, scientists/experts, the government and communities will allow each stakeholder to input their respective skills and knowledge, and enable the stakeholders to support each other, so as to achieve success.
- Different types of demonstration bases should be established according to the needs of the development. Examples include a bamboo forest management demonstration base, a new product development and innovation base, a bamboo product sales and market demonstration, bamboo forest ecotourism and ecological project demonstration, etc. The book provides a detailed introduction to the case of Lin'an County, Zhejiang Province, China, where different types of demonstration bases were established to guide the local bamboo shoot sector to achieve success.

5. Policies are the key factor for the sustainability of a bamboo sector. Practice has proved that correct policies are usually more important than financial investment. When a local government was able to issue different policies according to the needs of the development of the bamboo sector in different stages, the enterprises concerned could have a preferential environment to grow. Here, attention is drawn to the following points:

- Policy support should be systematic and continuous. Due to the comprehensiveness of the bamboo sector, a single policy may not be able to solve all problems, and so it is necessary to have a system of policies to support the sector, for example, the property and management rights of bamboo forests, land policies, taxes, finances, etc.
- The content and focus of the policies should vary at different stages of bamboo sector development, and adjustments need to be made to meet the specific needs of these different stages.
- When making decisions, the government should listen to and fully consider the needs and requirements of various stakeholders, and invite the stakeholders to participate in the decision-making process. The book introduces in detail the case of Anji County in Zhejiang Province, China, where the local government identified and implemented a series of policies at different stages of bamboo development, which guaranteed the sustainable and rapid development of the sector.

6. Capacity building and multi-participation. Special attention should be paid to capacity building, which should be considered to be a precondition for bamboo sector development. The following methods/approaches were identified as effective and as key based on experience gained from different locations of the world:

- Awareness raising. Before launching a bamboo development project, it is necessary to raise the awareness of various groups of people in the society concerned on the significance of bamboo development. At the same time, people should be helped to learn the basic conditions, technologies and skills that are necessary. The awareness of government officials, policy makers and investors should also be raised – an aspect that is usually neglected. From 1987 to 2015, I undertook a series of international training workshops, a major achievement of which was awareness raising for participants from different parts of the world. The participants were mainly composed of government officials, entrepreneurs and specialists. Their recognition of the significance of the bamboo sector and an understanding of how to develop the bamboo sector were the best outputs of the workshops. The book briefs on the impacts of MOST/INBAR training workshops, and takes the examples of Mr Romualdo L. Sta Ana, a Filipino participant, and of a Vietnamese government delegation, to illustrate the importance of awareness raising for pushing forward the development of the bamboo sector.
- Multi-participation and non-governmental organizations (NGOs). The development of a bamboo sector is closely related to the local economy, ecology and social development. It is necessary to have the participation of all stakeholders to guarantee the smooth and successful development of the sector. China and a number of countries with comparatively well-developed bamboo sectors have all established multi-participating NGOs, such as bamboo industry societies, bamboo

academic societies, bamboo foundations, cooperatives and research centres, etc. Most of these organizations have played positive roles in promoting their local bamboo sector. The book introduces the experiences of the Lin'an Bamboo Shoot Society in Zhejiang Province, China, and the Colombia Bamboo Society. Both organizations played multiple functions in their respective bamboo sectors and provided a platform for cooperation and exchanges among stakeholders; they also provided technical support and training for farmers and enterprises, and links with government, and made useful suggestions for decision making on policies and planning. They are also instrumental in organizing different stakeholders into common bodies to achieve win–win or multiple win purposes.

- Seek ways to establish multi-participatory and multiple win models. The promotion of harmonized, cooperative, mutually dependent – while still independent – and multiple win relationships among the stakeholders is a correct way towards achieving sustainability and success in bamboo sector development. The book introduces two company cooperative models, one a bamboo shoot processing enterprise – Kangxin Food Co. Ltd, and another a bamboo handicrafts enterprise – Yunhua Bamboo Tourism and Culture Co. Ltd. The model of 'company + farmer households + markets' did not only help the two companies to develop themselves, but also provided training to local farmers and thereby increased their income, as well as benefiting the local society and ecological environment. The model facilitated the sustainable development of the local bamboo sector and helped large number of farmers to find ways to get rid of poverty and achieve wealthy lives, as well as a win–win situation.

- The provision of expert consultancy at the initial stage of bamboo development, or when a special problem needs to be tackled, is usually necessary and should involve experts with rich practical experience and specific skills. Such experts will be able to help resolve the problems quickly and point out the right direction for the next steps. In the book, we introduce two case studies on this aspect. One is that of a young Chinese engineer who helped an Ecuadorian community to resolve the technical issues that they came across when Chinese machinery was first introduced into their facilities to process local *Guadua* and *Dendrocalamus asper*. The second case study is of a group of Chinese experts who provided detailed resolutions to and suggestions for a bamboo-processing company in Vietnam, after a thorough study on their major production and management problems. Adequate expert consultancy may be vital for the capacity building of an enterprise.

7. Market Development. The evolution of a bamboo sector usually sees a process of gradual development from low-end to high-end products. The processing industries develop from homestead to modernized, scaled-up processing, but no matter which development processing stage, market development is an everlasting issue that requires special attention. The construction and development of specialized bamboo raw materials market is very important. This book presents two case studies as examples: one is the development history of the Anji bamboo product market and raw materials market; another is how the Shilin Bamboo Product Co. Ltd has been developing its domestic and international markets, following on from its successful experiences in market development. At the same time, the book emphasizes the importance of farmer participation in product sales and marketing. Communities should be allowed to not only participate in raw bamboo production, but also in bamboo processing and product sales. Both of these case studies indicate that continuous innovation, and meeting the consumer's ever-changing needs are key to success in marketing. This is also the base for market development.

8. The bamboo product and sector life cycle and its related rules require special notice and investigation. Continuous innovation will inject energy into the sustainable development of the bamboo sector.

- Any bamboo product or sector may experience a developmental process from emergence to flourishing to decline, or from decline to flourishing again. This process happens almost everywhere and in almost every enterprise. The life cycle is a necessity, but sometimes it seemingly occurs by chance. However, there is always a reason for it. Examples include: changes in the quantity and quality of raw materials supply; changes in prices; changes in the labour market; changes in policies or in domestic and international markets; the development of new products and technologies; limitations in the management or hard/soft conditions of the enterprises concerned, etc. In order to provide detailed illustrations of the life cycle issue, we take as examples Japan and Taiwan, China, which were the earliest to develop their bamboo industries, and whose bamboo industries used to lead the world in bamboo development. The book gives a brief introduction to the histories of the rise and fall of these two bamboo sectors.

- Innovation is the power to drive sustainable development. The life cycle phenomenon exists, and cannot be avoided, but one should be prepared for any eventuality, and have continuous innovative thinking to meet emerging new challenges. An enterprise needs to continuously develop or access new technologies, new products and new facilities to adapt to the ever-changing market. At the same time, in order to meet its various challenges, bamboo enterprises need to seek ways to improve their management, inspire staff initiatives and improve staff quality; they need to continuously extend their supply chains and increase product value addition. In face of these challenges, local governments should develop new measures and issue new policies to provide a preferential environment for bamboo enterprises to develop. The book introduces one of China's pioneer bamboo companies – Dasso – describing its development story, and how the company has been growing and increasing in strength through its continuous technical revolution and product development.

The last chapter of the book emphasizes that all bamboo developers should develop their own featured bamboo sector or products according to specific local conditions and characteristics. This approach is also applicable to those countries, regions or enterprises that are just starting bamboo development. In this part of the book, the following points are stressed by the author:

- Specific strategies for bamboo sector development need to be identified according to local features. When identifying the local bamboo development strategy, an in-depth exploration should be made of the local existing bamboo resources, traditional craft and products, the ecological situation, and the development of policies and the social economy. Knowledge of these aspects will be very informative for any feasibility study. It is also necessary to study thoroughly the experiences of other countries, regions or enterprises, and identify suitable products and development models. Learning from the experiences of others does not mean simply copying everything – from products to models – but identifying certain principles of these experiences.
- Find the right breakthrough point. The right breakthrough point may be the first step towards success. In China, different bamboo-producing areas and enterprises have identified their own breakthrough points, and the same has been found in other countries. It takes extreme caution and deliberation to decide on a breakthrough point. The book discusses how Anji and Lin'an counties in Zhejiang Province and Yixing County in Guangdong Province in China, and the country of Vietnam, have successfully found their own breakthrough points.
- Attach importance to the strategic development of bamboo resources. Bamboo resources are the basis for identifying a bamboo sector development strategy. It is hard to utilize natural bamboo forests that have not been managed or rehabilitated for industrial purposes, so the rehabilitation and management of natural resources, and the development of plantations oriented towards industrial purposes, are very necessary for bamboo sector development. Evaluations should be made of the utilization of local bamboo resources, and an appropriate structure of bamboo species should be formed for further development.
- Plan before investment. I have already stressed and thoroughly discussed the importance of planning before investment. Here, I mainly discuss the principles of planning and the main contents of a bamboo development plan. The plan should not be invariable, but should be diversified and improved from time to time.
- Government should play an important role in the development of the bamboo sector, which involves many different government departments, encompasses many different parts of society and has impacts on many different stakeholders. It can be extremely hard to achieve sustainability of the bamboo sector when government support is missing, so the government should fully play leveraging roles for policies to push the bamboo sector to develop in a healthy way.

Financial tools should be involved when necessary. The government's role in the whole development process should be in providing services; it should avoid taking over the enterprises in making decisions. When bamboo enterprises meet challenges and difficulties, the government should provide all possible services and guides.

Zhu Zhaohua

Acknowledgements

The China Ministry of Science and Technology (MOST) has continuously supported international training workshops on bamboo topics for 17 years since 1999. MOST has also sponsored the author as the leader of Chinese expert groups visiting seven countries in Africa and Latin America from 2000 to 2003. At the invitation of the Prosperity Initiative (PI, a UK-based social enterprise comprising various donors, small enterprises, international businesses, farmers, processing companies and others) and the Ford Foundation, the author also worked with the Mekong Bamboo Project from 2003 to 2011. He was able to visit Vietnam nine times and Laos three times. Further, at the invitation of Mr Edgardo C. Manda, President of the Philippine Bamboo Foundation, the author visited the Philippines four times. At INBAR's invitation, he visited Ghana (2003), Ecuador, Colombia and Bolivia (2004), and India (2014). From 2006 to 2015, he was invited by various organizations to visit Bangladesh, Colombia, India, Myanmar, Nepal, Vietnam, Timor Leste, Nigeria, South Korea and Cameroon. The above-mentioned governments, international organizations, NGOs and enterprises provided the author with the opportunities to visit in person the major bamboo-producing countries in the world. The authors not only were able to providing consultancy services to, but also, taking advantage of the visits, were able to directly collect first-hand information on bamboo development in these countries. The many international workshops, study tours and short-term training courses held in China for different countries and organizations provided a platform for international exchange, and also opportunities for the authors to understand the experiences and challenges of bamboo development in different countries and regions, and to learn from peers.

Since 2003, the authors started to think about writing a book to give an account of successful and failed examples of development of the bamboo sector in China and in other countries. This thought was welcomed and supported by the then PI-sponsored Mekong Bamboo Project. From then on, the authors started to collect typical bamboo development case studies from a number of countries, with China studied as an important bamboo-producing country. With the support of many experts and friends, the authors were able to collect nearly 40 case studies. Here, the authors would like to especially thank the close cooperation and support of the following: Mr Chen Jianyin, Senior Forestry Engineer, Mr Chen Linquan, Deputy Director of the Forestry Bureau and Mr Xuan Taotao, Forestry Engineer of Anji County; and also of Mr Wang Anguo, Senior Forestry Engineer of Lin'an County and Mr Tang Mingrong, Deputy Director of Lin'an Forestry Bureau. These case studies have formed the basis of the book.

While collecting case study and other necessary materials for the book, the authors have also received friendly support and input from a number of experts in different regions:

- Prof. Ramesh Chandra Chaturvedi from India, who wrote section 2.5.5 The history of Indian bamboo utilization and development;
- Dr Ximena Londoño, who contributed 'Case 4: Introduction to the Colombia Bamboo Society' in section 3.6.2;
- Mr Romualdo L. Sta Ana, former President of the Philippine Bamboo Foundation, who came back to China nine times to attend international training workshops on bamboo, and who had been making efforts in pushing forward the development of the bamboo sector back in his home country. In section 3.6.1, under the subsection on 'Awareness raising for different groups of people – a case of MOST/INBAR international training programmes', he contributed to 'Case 1: Mr Romualdo L. Sta Ana – an alumnus from the Philippines';
- Prof. Shozo Shibata of Kyoto University, who provided detailed reference materials and his presentation on the 10th World Bamboo Congress, which made it possible for the authors to write 'Case 2: The evolution of Japan's bamboo industry' in section 3.8.3; and
- Prof. Moon Soontae, former Professor at Gwangju University, who provided materials and his paper on bamboo culture in South Korea for use in Chapter 2.

In addition, the authors have received support from a number of successful private entrepreneurs, who introduced the development history of their enterprises to the authors, and made it possible for their experiences to be shared with readers of the book across the world:

- Mr Chen Yunhua, President of Sichuan Qingshen Yunhua Bamboo Tourism and Culture Co. Ltd;
- Mr Wang Jianqin, President of Ningbo Shilin Arts and Crafts Co. Ltd;
- Mr Lin Hai, President of DASSO Industrial Group Co. Ltd; and
- Mr Germán Hugo Gutierrez-Céspedes and Manuella Mendes Araujo of the Industrial Group founded by João Santos, in Pernambuco, Brazil.
- Mr. Ye Ling, Director, Engineering Research Center for Bamboo Winding Composite Materials, China.

During the preparation of this book, the authors also received great help from the following people, who gave valuable suggestions for the book and in-time support when needed:

- Dr Ding Xingcui, China National Bamboo Development Center (CBRC);
- Mr Andrew Benton, former Manager of Outreach, INBAR;
- Mr Nelson O. Ononye, Chairman of Gamla Nigeria Co. Ltd., Nigeria;
- Mr Paul Vantomme, Senior Consultant at the Consultants Intérim SA, Belgium;
- Dr Li Yanxia, Senior Programme Officer, INBAR;
- Mr Jayaraman Durai, Project Manager, INBAR; and
- Dr Bai Yanfeng, Chinese Academy of Forestry (CAF).

Prof. Zhu Zhaohua would also like to thank Ms Su Juan, Ms Zhang Yan, Bhargavi Viswanath (India) and Ms Gan Huimin, who provided translation, preliminary editing and data sorting assistance, as well as a number of experts from different parts of the world who provided precious reference materials and photos.

To sum up, the book would have been impossible without the support and help of the above-mentioned organizations and people.

1

Introduction

1.1 The Imbalanced Status of Bamboo Development in the World

Bamboo is widely distributed in the developing countries of Africa, Asia and Latin America, while many developed countries in Europe and North America, including Japan and Australia, are key consumer countries of bamboo products. Among these developed countries, Japan is an exception, because it has a natural distribution of bamboos. The country used to have a very developed bamboo sector, but because of labour shortage and costs, there is no longer any large-scale bamboo production in Japan, and it has now become an important bamboo consumer country, especially of bamboo food products.

Although the global distribution of bamboos is quite wide, people's awareness of the roles of bamboo is quite different. Before the 1980s, in most bamboo-producing countries, bamboos were still growing in natural stands, with little or no management. Bamboo products were made in a traditional, handmade way, using traditional technologies, and they were also traditional products for local markets; there was little or no industrial processing.

However, earlier, before the 1950s, some bamboo-producing countries, influenced by their long history of civilization and traditions with bamboo, had started research on bamboo biodiversity, its ecological and biological characteristics, its timber properties and processing technologies.

For example, India, China, Brazil and Vietnam developed bamboo pulp processing industries. Equally, Colombia and some other Latin American countries established bamboo construction industries. In China and other South-east Asian countries, a small-scale bamboo shoot processing industry already had a comparatively long history. In some Latin American countries, such as Colombia, Ecuador and Peru, and in some Asian countries, such as the Philippines and China, the bamboo furniture industry was also well developed before the 1980s. Between the 1950s and 1960s, China and India started to research and develop ply bamboo and boards. At the same time, Japan and Taiwan, China (henceforward 'Taiwan') had started the mechanical processing of bamboo mats, curtains, sticks and laminated boards. From 1986, Mainland China began to introduce bamboo-processing machinery from Taiwan and thus a larger scale and industrialized bamboo sector came into being. Not long after this, it was possible to produce all types of bamboo-processing equipment in Mainland China.

Until recently, there were still development gaps and an imbalanced development situation among bamboo-producing countries. However, in the 1980s, under the influence of a number of early bamboo-developing countries, the first worldwide non-governmental organization (NGO) concerning bamboo – the World Bamboo Organization (WBO) – was established (in 1984), and organized the First World Bamboo Congress

Fig. 1.1. Bamboo fossil from the 2014 Beijing Garden Expo (Zhu Zhaohua).

(WBC). In 1997, the first intergovernmental bamboo and rattan network – the International Network for Bamboo and Rattan (INBAR; since May 2017 the International Bamboo and Rattan Organization) was launched. As a result of the efforts of the above organizations, global awareness of bamboos and bamboo development, and their roles, were greatly raised.

1.2   The Co-efforts of the IDRC, IFAD and a Number of Countries in Promoting Bamboo Sectors in the World, and the Establishment of INBAR

> *The first to introduce bamboo to international society, the IDRC has made crucial contributions.*

The International Development Research Centre (IDRC) of Canada was the first organization that introduced bamboo into global view, and raised attention of it in the international sphere. IDRC sponsored an international workshop on bamboo and rattan in Singapore in 1980, and this initiated international exchange and communication about these two valuable non-timber forest resources that used to be neglected. From then on, IDRC started to sponsor major producing countries of these two resources in conducting research. This research has led to a wide interest, with the number of participating countries and organizations increasing rapidly. In 1993, IDRC organized an international conference in India, where INBAR was formed as an international cooperation and development programme. In 1994, INBAR succeeded in obtaining funding from the International Fund for Agricultural Development (IFAD). The early projects of INBAR successfully raised people's awareness of the significance of protecting and developing the two important plants and their related industries. However, people soon realized that a single international cooperative programme may not be able to continuously support the promotion of global bamboo and rattan sectors in the long run.

In order to promote the protection of biodiversity and the sustainable development of bamboo and rattan globally, and improve their services in the construction of ecological systems and contributions to livelihood improvement in poor regions and countries – as well as their benefits to consumers and producers – IDRC, IFAD, and a number of major countries participating in the INBAR programme (including China, India, Malaysia, Thailand, etc.), started discussions about establishing an independent and permanent international organization based on the original INBAR programme. In March 1995, consultants of the INBAR programme and the main project experts from IDRC, IFAD, India, China, Malaysia and Thailand held a meeting in Malaysia, after which a Special Taskforce Group for the Preparation of INBAR Internationalization was established, and it was decided that INBAR's headquarters would be located in China. Soon after this, China established the 'China Leading Group of INBAR Launching', which was composed of the representatives from the Chinese Ministry of Foreign Affairs, Ministry of Science and Technology, Ministry of Forestry, Ministry of Finance and the Municipal Government of Beijing. Over more than 3 years of cooperation, INBAR was established as an independent, intergovernmental and non-profit international organization on 7 November 1997 in Beijing. This INBAR is different from the INBAR mentioned earlier (which is an international programme initiated in 1993), although it shares the same name, and it works as the first international organization dealing with bamboo and rattan. At its launch, INBAR had nine founding member countries; by 2017, the number of member countries had increased to 43; except for Canada, all of the member countries are bamboo and rattan-producing countries located in Asia, Africa, Latin America, the Caribbean and Oceania.

According to its establishment Treaty, the Mission of INBAR is 'to improve the well-being of producers and users of bamboo and rattan within the context of a sustainable bamboo and rattan resource base by consolidating, coordinating and supporting strategic and adaptive research and development'. Since its establishment, INBAR has carried out a series of global and regional activities, and fostered cooperation with and among member countries; it has also played the role of an exchange network and platform for bamboo and rattan development strategies, policies, scientific research, technology transfer and international training (see INBAR, 2007). Professor Zhu was honoured to have been involved in the whole processes of preparation and launch of INBAR.

1.3 The WBO and WBC as Important International Cooperation and Exchange Platforms

1.3.1 Mission and goals of the World Bamboo Organization (WBO)

The WBO is a diverse group consisting of individual people, commercial businesses, non-profit associations, institutions and allied trade corporations that all share a common interest in bamboo. Its purpose is to improve and promote bamboo, as well as the conditions affecting it and the industry surrounding it. The Organization is dedicated to promoting the use of bamboo and bamboo products for the sake of the environment and economy.

The WBO is a US tax-exempt trade association that was formed to facilitate the exchange of information from around the world on the environmental, socio-economic, biological and cultural aspects of bamboo. By bringing together people concerned with bamboo and creating mechanisms for global communication, the WBO's goal is to facilitate the development of partnerships and alliances to advance the cause of bamboo worldwide.

1.3.2 The World Bamboo Congress (WBC)

The WBC is a unique event that encourages global interaction by providing a platform for direct networking and the sharing of ideas and information. Up to 2015, ten WBCs were held: in Puerto Rico (1984), France (1988), Thailand (1991), Indonesia (1995), Costa Rica (1998), India (2004), Thailand (2009), Belgium (2012) and South Korea (2015).

The Congress has become an important international exchange platform on bamboo issues. Each Congress has attracted large numbers of bamboo experts, who gathered to exchange information on the latest research results and new products. The theme of each Congress varies according to the status of bamboo development in the world, and the 'hot' topics have been ever-changing. The main issues that have been the concern of the Congress have been guiding the world bamboo sector by providing new fields and directions of development.

Below is a simple introduction to the Tenth WBC held in Damyang, (South) Korea, in 2015 (WBC, 2015), which describes the activities that were involved.

The World Bamboo Fair

The Fair was divided into two parts, the Korea Pavilion and the International Pavilion. Rich information and many fine products were found in the Korea Pavilion. The exhibition gave a systematic introduction to the development history, resources, management and culture of bamboo in the country. The bamboo industry of South Korea features fine and high value-added handicrafts, for example, bamboo musical instruments and fine crafts. Besides these products, South Korea has also developed innovative bamboo-processing techniques for cosmetics, health drinks, medicines, special bamboo salts, bamboo composite materials and bamboo coffee and wines, etc. The International Pavilion gathered various bamboo products from all over the world, ranging from those most advanced in technology to the most special in cultural features.

The Bamboo Seminar

The Seminar was held on 18–22 September 2015, and was a very compact but ordered event. The contents were divided into topics on bamboo architecture, bamboo and climate change, bamboo and environment, large-scale bamboo silviculture, bamboo and community development, etc. In the Seminar, bamboo architecture and bamboo carbon trade seemed to be the main focus, and there were quite a number of articles and presentations on architecture. These presentations were of wide interest to the participants from Europe, North America, Asia, Latin America and Africa. More than 400 participants from around 30 countries attended the Seminar.

World Bamboo Day

At the Eighth WBC (2009) in Thailand, the Thai Royal Forestry Department proposed the designation of September 18 of each year as World Bamboo Day, and more than 350 participants of the Congress from 41 countries agreed to this initiative. Representatives from each country planted a bamboo plantlet at the Garden of the Thai Royal Research and Development Center as a token of their respective countries. Since then, 18 September each year has become an opportunity to speak out loudly about bamboos, plant bamboos and carry out other bamboo-related activities. The initiative has successfully promoted the protection of bamboo resources, as well as their sustainable management and utilization worldwide.

The World Bamboo Design Competition

In 2014, the Organizing Committee of the WBC decided to hold a 'World Bamboo Design Competition' during the 2015 Congress. This event was supported by the Korea Institute of Design Promotion, the Forest Service and the Ministry of Agriculture, Food and Rural Affairs. Winners included designers from India, Portugal and the Netherlands for the Household Goods Category, from South Korea, China, Ghana, the Netherlands and Vietnam for the Transportation Category and from South Korea, Indonesia and the

Philippines for the Architecture Category. The prizes ranged from US$1000 to 10,000.

The World Bamboo Congress Bamboo Pioneer Award

The WBC Bamboo Pioneer Award was first initiated in the Eighth WBC in Bangkok, Thailand. Why this award? Its introduction as stated by the WBO was:

> From its reputation as a 'poor man's timber' to its current potential as a high-end product that provides better structural, architectural and visual qualities over traditional alternatives, bamboo's use globally has progressed at an unprecedented rate. Additionally, bamboo has been rediscovered as an alternative fibre, an alternative to fossil fuels, a substitute for plastic, a nutritional food supplement, and a green resource to mitigate climate change, as well a viable tool for rural economic development. Dedication, determination and collaboration are required to advance any scientific endeavour, and those individuals whose lifelong commitment to bamboo science deserve our attention and honoured recognition.

At the WBC in Thailand, nine people won the awards; they were from Japan, Thailand, the USA, Germany, Canada, Taiwan, Colombia, the Netherlands and Mainland China. At the Damyang WBC, four people won the awards. They were Professor Jorge Morán from Ecuador, who has been researching on bamboo architecture; Dr Dina Nath Tewari from India, who has been working on rural development and relative policy studies; Mr Choi Hyungsik from South Korea, Governor of Damyang County, who has been promoting the development of effective use of bamboo in the county, and was also a great contributor to the Damyang WBC. Professor Zhu was honoured when awarded the title of Bamboo Pioneer at the Damyang WBC.

Launch of the World Bamboo Ambassadors platform

In order to facilitate international cooperation on bamboo development, and to raise people's awareness of the status of bamboo development in different countries and regions, the WBC launched a new platform in 2015 called the 'World Bamboo Ambassadors'. The Ambassadors

Fig. 1.2. The World Bamboo Pioneer Award at the 2015 World Bamboo Congress (from left to right: Michel Abadie, President of the WBO; Jorge Moran; Choi Hyungsi; Susanne Lucas; D.N. Tewari; and Zhu Zhaohua).

comprised a number of enthusiastic, progressive and dedicated persons from important bamboo-producing countries who wanted to push forward the global potential of bamboo.

1.4   The Impacts of International Training Activities in China on World Awareness of Bamboo and its Roles

The special roles of China in raising the world's awareness on bamboo

With the support of China's Ministry of Science and Technology (MOST) and its Ministry of Commerce (MOFCOM), INBAR has cooperated with the International Farm Forestry Training Center (INFORTRACE) of the Chinese Academy of Forestry (CAF), the Lin'an Modern Forestry Technology Service Center (LMFTSC) in Zhejiang Province, the China National Bamboo Research Center (CBRC) of the State Forestry Administration (SFA) and the China International Center for Bamboo and Rattan (ICBR) in organizing a series of international training activities on bamboo and rattan. Now, let me take the earliest international bamboo training courses

as an example – the MOST/INBAR annual training programme. From 1998 to 2015, more than 1100 people were trained in the programme, of which a major part of the participants came from developing countries in Asia, Africa, Latin America and the Caribbean and Oceania, with a lesser part coming from Europe, North America, Japan and related international organizations. In addition, at the request of member countries and related international organizations, the training programme has also organized in total 25 special study tours and training courses. Zhu Zhaohua and Jin Wei played major roles in organizing the above training courses and study tours.

The above capacity-building activities have significantly raised world awareness of the impact of the roles of bamboo and rattan in ecological construction and poverty alleviation, and international society has accepted the concept that bamboo and rattan products are environmentally friendly. At the same time, a very important role of these training activities was to provide a platform for the exchange of experiences and technology transfer among countries. The above achievements have resulted in a great increase of interest among bamboo-producing countries. A number of countries have started to study China's industrial bamboo-processing technologies, introduce machinery from China's mainland and Taiwan area, and initiate the industrialization of their own bamboo sectors.

1.5   The Development of Bamboo Industries in Bamboo-producing Regions

Through two decades years of practice, we can see a number of successful cases in the industrialization of the bamboo sector. In Asia, bamboo development is faster and has great advantages in many aspects compared with other parts of the world. India's bamboo industrialization is now on full course, and there have been successful cases in the development of bamboo panels, the use of bamboo in construction and bamboo energy forests. In the Philippines, many government officials, entrepreneurs and researchers have attached great importance to the development of bamboo; they have been working actively in the preparation of large-area plantations, and have been quite innovative in bamboo construction, furniture and engineered board production. Vietnam has introduced more than 100 bamboo primary-processing production lines and equipment from China, and has been very successful in making bamboo chopsticks, bamboo sticks and laminated boards, and in engineered board production. Vietnam has also introduced equipment for the production of Pressed bamboo materials (other names describing the same material include: bamboo-based fiber composite, bamboo scrimber, strand woven bamboo material, reconstituted bamboo material, etc., herein after all called pressed bamboo) and is now able to produce pressed bamboo on a large scale, and to further process these materials into flooring, decorative boards and furniture. Although there are still some problems, such as low utilization rate and low product quality, Vietnam has basically been successful in the industrial development of its bamboo sector. The country has been especially successful in the industrial planting and processing of sympodial bamboos.

In Africa, countries such as Ghana and Ethiopia have already started the industrial processing of bamboo curtains and charcoal, while Rwanda, Ethiopia, Madagascar, Ghana and Kenya have started to develop bamboo plantations for culm and shoot production. In Latin America and the Caribbean, Ecuador, Colombia, Peru, Bolivia, Brazil, Mexico and Costa Rica, there has been further development of their bamboo construction sectors, and positive results have been achieved in the production of bamboo laminated board and furniture. Chile and Ecuador, respectively, introduced fine sympodial and monopodial bamboo species from China in 2008 and 2003. Chile has made great breakthroughs in the processing of shoots of *Chusquea culeou*, and the canned shoots were welcomed in the market. Brazil, Ecuador, Colombia, Peru and Mexico have started to attach importance to the development of bamboo plantations. Brazil especially has established large-scale industrial plantations for pulp purposes, and has successfully managed the plantations for 40 years.

1.6   Problems in Bamboo Industry Development Still Exist in Some Countries and Regions

Through the efforts of the international organizations introduced above – such as WBO, INBAR, and a series of international training activities carried out in China, as well as the efforts of a number of bamboo-producing countries, the world's awareness of bamboo-based development has been greatly improved. However, it is worth particular notice that there are still quite a number of bamboo-producing countries in the world, including many countries in Africa, where bamboos are not getting wide attention, and are still left in the wild without management. In some countries, bamboos are even treated like weeds that need to be eliminated. Many countries still lack clear policies on bamboo resource management and utilization, and because there are no regulations on property ownership, the bamboos can be harvested at free will. Some other countries have ended up doing the opposite – they have banned natural bamboo forest harvesting, and thus, without reasonable management and utilization, natural bamboo forests are left to degrade over large areas. In other countries, it can be seen that after realization of the values of bamboo and bamboo development – especially by entrepreneurs and citizens of those countries, and inspired by their visits and studies in China – the purchase of equipment and building of factories has been commenced hurriedly, without a careful feasibility study and strategic plan, and so the projects have finally failed. There have been quite a number of such cases.

Based on the (first) author's observations and studies on the cases of 23 bamboo-producing countries in Asia, Africa and Latin America, and on information gathered from participants in international training workshops and seminars, combined with his research on China's bamboo development history, he feels deeply that developing a bamboo sector is quite complex. It involves stakeholders from a wide range of working fields and different walks of life; at the same time, there are a number of key factors that affect the development course taken by the sector. The development of a bamboo sector is not simply related to technical and investment issues, and the industrial production of bamboo is more complicated and difficult than other 'normal' production processes. The basics of a bamboo production industry include large-scale, highly qualified bamboo resources and a favourable policy and investment environment; at the same time, the development of the industry needs a thorough strategic plan that has carefully considered the local conditions and the practical feasibility. Other must-have conditions for achieving sustainability and success include the multi-participation of the government, private sectors, bamboo producers and scientists, which is key for realizing a multiple win situation, and the continuous innovation of new products that meet market demands.

Although the development of a bamboo sector is very complicated, a successful and sustainable bamboo sector can be attained through an understanding of the development rules of the bamboo sector, by providing the necessary basic conditions that the development of the sector needs, and by multi-stakeholder participation. Most important is the development of a featured bamboo sector that will fit into the special conditions of the country or the region, while learning from the good practical experiences of the other countries and regions.

1.7 The Purposes of the Book

In order to achieve sustainability of the bamboo sector in various regions and countries, and

under varying conditions, there are two basic points that we want to note in the international training projects we are organizing in collaboration with INBAR and MOST. One is to introduce China's successful experiences and practices to other countries. The other is to increase awareness that holding training courses in China does not mean that the delegates have to duplicate China's development mode and products at home; on the contrary, it means exploring ways of development that meet the special conditions of different countries and regions by investigating valuable experiences from China that have universal significance, and combining these with the specific development orientations, strategic approaches and product strategies that are identified based on local bamboo resources, and social, economic and personnel conditions. Through analysing the successful and failed cases in China and a number of other bamboo-producing countries in the world, in combination with studies made in 23 countries, knowledge learned in bamboo-related international training workshops, and study tours to China organized for a number of countries and related international organizations, and participation in international seminars, the authors would like to share and discuss with the readers of the book our learning from experience, in the hope that this piece of work inspires more valuable and remarkable cogitation and initiatives on the sustainable and healthy development of the world's bamboo sector.

In this book, there are three primary questions the authors would like to answer and discuss with readers. First, compared with other plants, especially trees, what are the special features of bamboo and its outstanding contribution to human beings? Second, a question concerning most people and a motive that led me to write this book: how may the bamboo industry be developed in a sustainable way? We try to bring some successful practices and lessons learnt from failure cases around the world, especially practices and experiences from China's bamboo sector, to elaborate on eight factors that affect the sustainable development of the bamboo industry, factors that the authors believe should be valued during development of the bamboo industry. Third, what are my humble suggestions to countries and enterprises that have just started their bamboo businesses. In addition to posing these questions, this book has been inspired and facilitated by many of the authors' peers working internationally who have participated in my training workshops or interviews, and want Professor Zhu to officially publish a book to share his knowledge, experiences and understanding. Professor Zhu considers this book to be a modest inspiration for all friends who are interested in the sustainable development of bamboos.

References

INBAR (2007) *In Partnership for a Better World – 15th Anniversary of the International Network for Bamboo and Rattan*. International Network for Bamboo and Rattan, Beijing.

WBC (2015) *Conference Guide: 10th World Bamboo Congress, 17 Sept–22 Sept 2015, Damyang, Korea*. World Bamboo Congress, Plymouth, Massachusetts. Welcome document available at: http://www.worldbamboo.net/wbcx/Welcome.pdf (accessed 15 June 2017).

2

The Contribution of Bamboo to Human Beings Is Far More Than Is Imagined

---

2.1 Bamboo is One of the Easiest to be Sustainably Managed (Self-renewable) Plants

2.1.1 Is bamboo a tree or a grass?

Bamboos are a special type of plant. They have a tenacity that trees cannot compare with and special properties that make them very easily sliced; both are strong in self-renewal. At the same time, bamboos have the characteristics that herbaceous plants do not possess, such as high hardness and elasticity; they can maintain exuberant growing capacity for many years. The international community classifies bamboo as a non-timber forest resource. China's first monograph about bamboo was called *The Bamboo Monograph* (*Zhu Pu*, 《竹谱》). It was written by Dai Kaizhi (戴凯之) during the Southern Dynasties (AD 420–589). The first line of the book says 'Among all the plants, there is one called bamboo, the property is not too strong nor too soft, it belongs to neither herb nor wood (植类之中，有物约竹，不刚不柔，非草非木。).' Another line says: 'If bamboo is a kind of herb, then it should not be named as bamboo, thus, none of the herbs may perform like bamboo. … bamboo is a name for one type of plants, or another name for a category of plants that have the same properties, among the plants there are herbs, trees and bamboos, just like among the animals there are

fish, birds and beasts (若谓竹是草，不应称竹，今既称竹，则非草可矣……竹是一族之总名，一形之偏称也,植物之中有草、木、竹,犹动物之中有鱼、鸟、兽也。).'

In recent years, there has been a consensus that the bamboo is herbaceous; some international publications have also identified it as a herbaceous plant, for example, in some Indian publications, bamboos are recognized as herbaceous (Ramanuja Rao, 2013). However, I (herein after refer to the first Author – Prof. Zhu Zhaohua) consider that it is more accurate to classify it as a special species that is neither herbaceous nor arboreal. This, of course, does not include a small number of herbaceous bamboo species in Latin America and Africa, because more than 92% of the world's bamboo species are woody bamboos. This classification of bamboo is based on the ecological and biological characteristics of bamboos, as well as in recognition of their properties and features in utilization. These characteristics, properties and features make them identifiable from timber species and herbs, which is essential for the development of the bamboo industry in modern times, because it helps to stop people from applying the same concepts and modes of timber tree management to bamboos. For example, some countries engage in bamboo forest certification, which is not in conformity with the ecological and biological characteristics of bamboos, and hinders further research on bamboos. We advocate using bamboo to replace

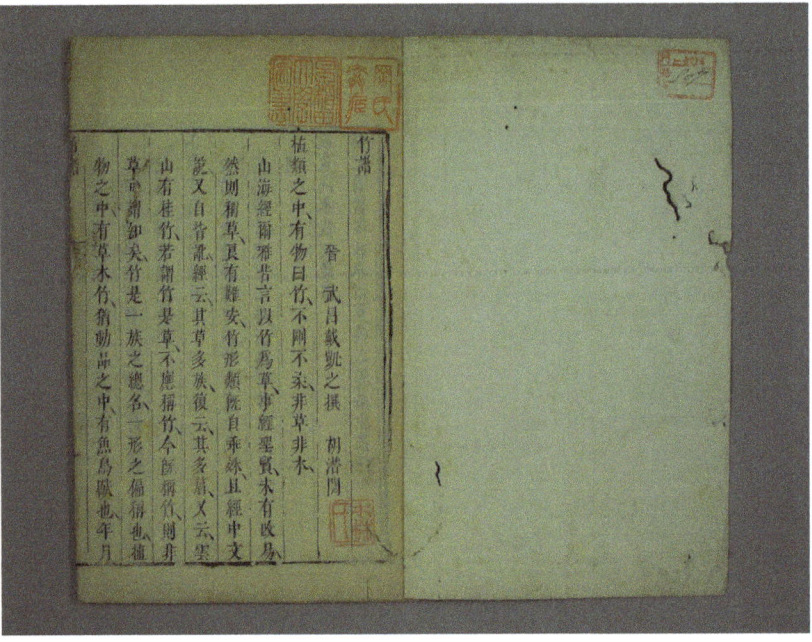

Fig. 2.1. Pages from *The Bamboo Monograph* (*Zhu Pu*, 《竹谱》), written by Dai Kaizhi (戴凯之) during the Southern Dynasties (AD 420–589).

(tree) wood, in the hope of giving full play to the advantages of bamboo plants in protecting the environment, preserving forest and meeting human needs. Besides, the distinction between the two types of timber helps to avoid market policy on (tree) wood products from applying to bamboo products. Now the international community tends to recognize bamboo and bamboo products as non-wood forest products, which is a very scientific understanding and conducive to the development of the bamboo industry.

2.1.2   Diversity and distribution of bamboos

Recent studies show that there are 1482 bamboo species in the world.

According to Ohrnberger and Goerrings (1985; cited in Seethalakshmi and Muktesh Kumar, 1998) there are 110 genera of bamboo recorded in the world and 1110–1140 species. Since 1985, there have been many more in-depth investigations on bamboo and a lot of new findings. For example, in Vol. 22 of the *Flora of China* (Wu Zhengyi and Raven, 2006), the discovery of 34 genera and 534 species in the following taxonomic unit was recorded: 440 subspecies, 36 varieties, 68 variants and four artificial hybrids. In 2008, Yi Tongpei, Editor in Chief of the *Chinese Bamboo Annals*, recorded 43 genera, 707 species, 52 varieties, 98 variants and a total of 857 species and subspecies (Yi Tongpei, 2008). From 1996 to 2007 there were more than 200 new species recorded.

According to Emmett J. Judziewicz, Ximena Londoño and Lynn G. Clark, who published *American Bamboos* (Judziewicz *et al.*, 1999), the Americas have a record of 430 species of woody bamboo, of which 40% belong to *Chusquea* species. In 2015, when I met Ximena Londoño, one of the authors mentioned above, I was told that in the Americas, the bamboo species described have increased to 531, a significant rise compared with the numbers in 1999.

B.N. Kigomo's *Distribution, Cultivation and Research Status of Bamboo in Eastern Africa* (Kigomo, 1988), recorded that Africa has 43 species and 14 genera, of which only three genera and three species are distributed in the mainland. The other 11 genera and 40 species are distributed in Madagascar. The book also records the distribution of two kinds of tall reed grass in the African continent. In 2015, according to L.G. Clark, X. Londoño and E. Ruiz-Sanchez, there are 1482 described

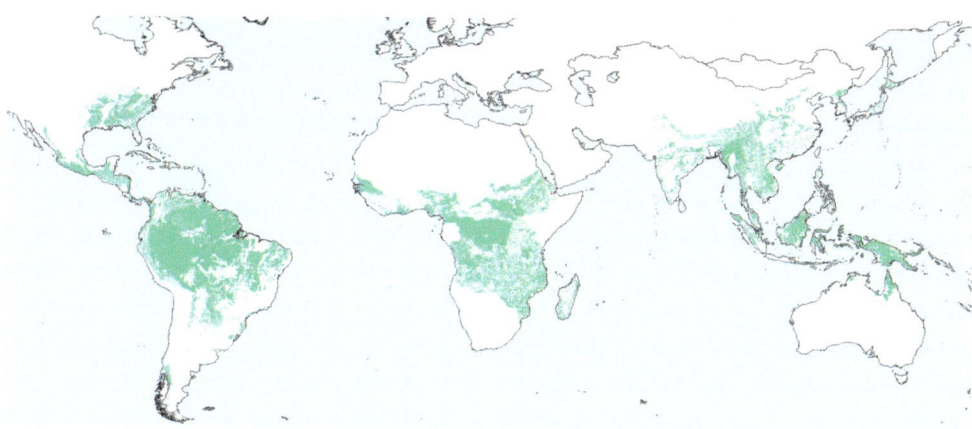

Fig. 2.2. Potential distribution of bamboos in forest areas in the world (INBAR).

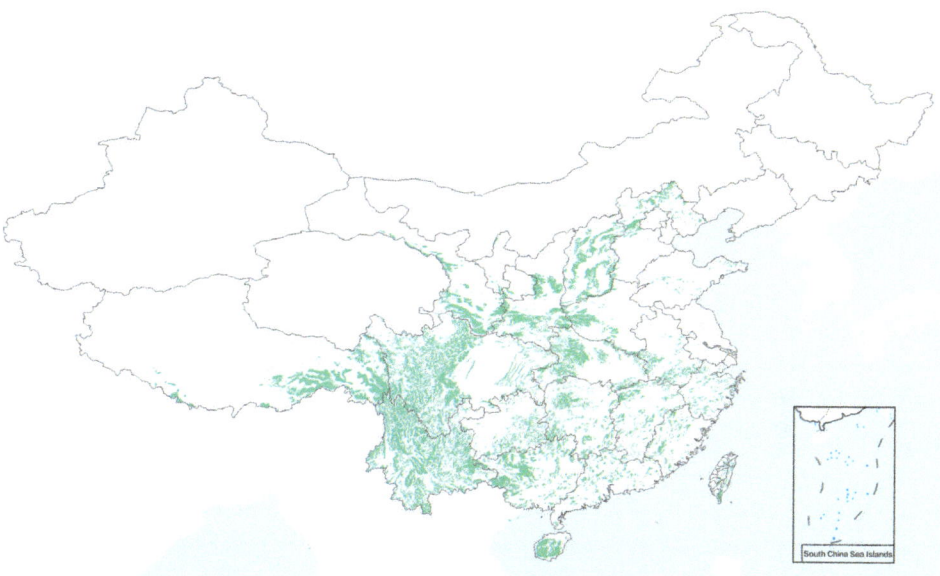

Fig. 2.3. Potential distribution of bamboos in forest areas in China (INBAR).

species of bamboo and classified into 119 genera and three tribes (Clark *et al.*, 2015): *Arundinarieae* (temperate woody bamboos, 546 species), *Bambuseae* (tropical woody bamboos, 812 species), and *Olyreae* (herbaceous bamboos, 124 species). The total number of bamboo species is ever changing, as there are always new findings and developments in the taxonomy of bamboos, and the diversified methods and ways of recognizing the species and naming them. Before the completion of the book, the authors were made aware in a newly published book, that the number of bamboo species in the world was updated to 1642. (M.S. Vorontsova *et al.*, 2016).

The herbaceous bamboos are mostly distributed in the Americas, with more than 100 species; 119 genera of bamboos are native to all continents except North America and Europe, and have a latitudinal distribution from 50°30′N to 47°S and an altitudinal distribution from 0 to 4300 m. Therefore, it is believed that there are two distribution centres of bamboo: one is in the southern and southwestern part of China, South-east Asia and the eastern part of South Asia, while the other is located in South America. Asia has 900 to 1000 species belonging to 70 to 80 genera.

2.1.3 Propagation and regeneration characteristics of bamboos

Bamboos can be propagated either by using vegetative methods, such as underground stems (rhizomes), culms and branches or by seeds. Though the propagation techniques are quite varied, they are relatively easy to master. The stem of bamboo has a notable morphological characteristic – it has underground stems and above-ground stems. The underground stems are what we usually called the rhizomes. The above-ground stems are the culms; they have branches and leaves. The rhizomes are actually the main stems of bamboo, while the culms above the ground are actually secondary stems. Bamboos can be divided into three categories according to the morphological characteristics of their rhizomes: monopodial rhizomes (known as 'running'), sympodial rhizomes (known as 'clustered') and amphipodial rhizomes (known as 'compound'). Each category has some variations and there are no 'typical' morphological characteristics (see Fig. 2.10).

Research on the propagation technologies (especially vegetative propagation) of bamboos is mainly focused on whether the rhizome buds or culm buds can be actively grown into new plants and under what conditions (temperature, humidity, bud development). Currently, the clone methods are the dominant means for bamboo propagation. Nowadays, quite a number of bamboo species have been successfully reproduced through tissue culture methods, especially sympodial bamboos. However, up-to-date, tissue culture methods are still not successful for most of the monopodial bamboos, and their propagation still needs to rely on the transplanting of mother stock or rhizomes. Although the propagation of seed works well, we cannot completely depend on that. Most bamboos do not flower annually, there are no methods of predicting the flowering time of the bamboos and the mechanism of bamboo flowering is still a mystery to us. Fortunately, through long-term research on the biological characteristics of bamboos, now we can use different propagation methods for different bamboo species, and bamboos can be propagated at low cost and high efficiency.

Fig. 2.4. *Glaziophyton mirabile*, Bai Coroado – bamboo without leaves (Ximena Londoño).

*Bamboo has high natural
self-renewable abilities*

A study by China Southwest Forestry University (Yang Yuming and Sun Maosheng, 2005) revealed that a cluster of *Dendrocalamus giganteus*

Fig. 2.5. *Neurolepis mollis* 993 XL 82ER 011 – bamboo of the largest leaves (Ximena Londoño).

with seven to nine bamboo culms can generate 591 new bamboo shoots in a year. Under normal management and natural conditions, 51% of these new shoots will mature; while under better site conditions, and with intensive management and rational shoot thinning, the maturation rate will greatly be increased. Take *Phyllostachys heterocycla* var. *pubescens* (Moso bamboo), for example: in a plantation established in Yixing Forest Farm, Jiangsu Province, China in February 1959, the average number of new culms generated by one mother bamboo was 168 in only 7 years (up to 1965). The above features of bamboo growth are the most appreciated and welcomed – once planted, bamboo can self-reproduce every year, and the plantation can be partly harvested once a year to bring in more income. The plantation can also help to maintain a sustained and stable ecological system. Bamboos are among the strongest woody plants in self-renewal.

2.2 Bamboos Are Fast-growing with High Productivity and Economic Value

Amazing! Bamboo can grow to 20 m high in only about 50 days!

Fig. 2.6. A bamboo nursery – propagation by culm cuttings (Zhu Zhaohua).

Fig. 2.7. Bamboo plantlets grown from culm cuttings at 40 days old (Zhu Zhaohua).

Fig. 2.8a,b. Bamboo tissue cultured in the greenhouse (Zhu Zhaohua).

2.2.1 Fast growth is one of the notable characteristics of bamboo

In China, when describing something such as a new emerging industry which is developing fast, people say that it is 'like spring shoots after rain', though when it is underground, the lateral buds take a long time to emerge from the rhizome and grow into bamboo shoots. Take Moso bamboo for instance; the lateral buds usually sprout and become shoot buds in late summer and early autumn, and then shoot buds shape into bamboo shoots by early winter. After dormancy in the cold winter, the shoot buds continue to grow rapidly into new bamboo shoots underground until they emerge above ground in early spring. This process takes 3–6 months from lateral bud sprouting to growth of the bamboo shoots underground, while it only takes 45–55 days from the bamboo shoots emerging from the ground surface to reaching their full height of 7–10 m. The time for bamboo shoot sprouting varies with the species and season. For the tropical sympodial bamboo species, such as *Dendrocalamus giganteus*, which is on average 20 m high, diameter at breast height (DBH) is around

Fig. 2.9. A bamboo nursery garden with *Indocalamus* spp. propagated by rhizome cuttings (Jin Wei).

20 cm, and it takes only 50–60 days for the lateral buds to differentiate and become bamboo shoot buds, and grow into bamboo shoots emerging from the ground; it then takes another 40–50 days for the new shoots to finish their height growth. The speed of growth is amazingly high. According to research, *Pleioblastus amarus*, a monopodial species, can grow to 1.21 m tall within only a day (24 h). I often get questions from foreign friends who are unfamiliar with bamboos, such as 'How long does it take for a bamboo to grow into this size?' My answer is 'about 50 days'. Such replies always receive gasps of amazement. In Fig. 2.17, we can see that bamboos only need about 2 months to grow to their final height, and that they then keep the same size (height and DBH) until death.

2.2.2 Fast maturation of a bamboo forest

Bamboo plantations can mature into productive bamboo forests in a comparatively short period of time. Moso bamboo needs 8–10 years to mature, which is the longest time needed among all bamboo species. For forests raised for shoot purposes, e.g. *Phyllostachys praecox*, it takes only

4 years for the stand to mature and become productive, while sympodial bamboo species need 3–5 years.

2.2.3 High biomass production by bamboo forest

The vigorous growth of a bamboo forest is retained with an annual selective harvesting of 20–25%. Reasonable harvesting does not harm the normal growth of bamboo forest and conserves the capacity for soil conservation; at the same time, it improves the sustainability and unit yields of the forest. According to the research of Prof. Ma Naixun (Ma Naixun, 2001), the biomass of bamboo culms is in general relatively high, although it varies with species (see Table 2.1). Bamboo forests not only provide bamboo culms every year, but also produce large amounts of edible shoots as food, with which trees can hardly compare. In India, Grow More Bio-Tech Ltd has planted 631 ha of *Bambusa balcooa* for bioenergy; the average annual biomass above-ground is about 100 t/ha, and the total aboveground biomass can accumulate to 926 t/ha over 10 years.

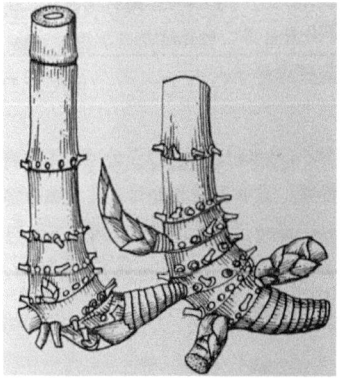
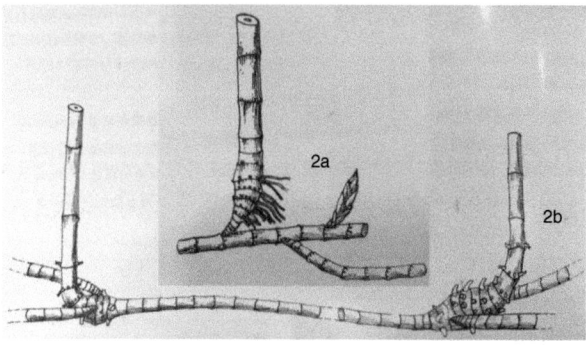
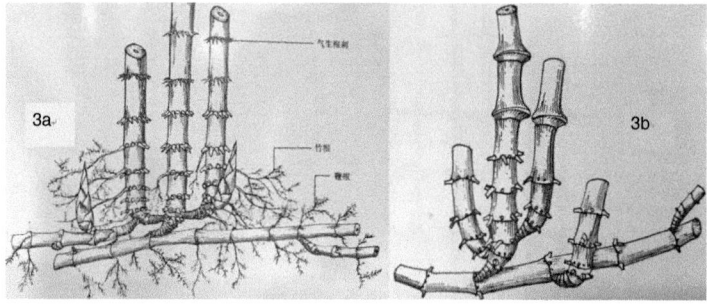

Fig. 2.10. Rhizome patterns in American woody bamboos. Root parts of sympodial (top), monopodial (middle) and amphipodial (bottom) bamboos (from Judziewicz *et al.*, 1999, p. 8, with permission from the authors).

2.2.4 The high economic value of bamboo

Every part of bamboo is a treasure with high economic value

Taking Moso bamboo as an example, the culms, leaves, branches, rhizomes and roots can all be processed into many different products. The income from bamboo shoots can sometimes exceed the total income from culms, branches and leaves. Many years of practice in Anji County, Zhejiang Province, show that the eco-nomic benefits from pure-culm-oriented forests are always lower than those from culm-and-shoot-oriented forests or pure shoot-oriented forests (see Table 2.2) (Zhu Zhaohua, 2013).

Case studies

CASE STUDY 1: COMPARISON OF THE ECONOMIC BENEFIT OF BAMBOO FOREST AND CROPS. Professor Zhu conducted an investigation in 2003, and the results showed that the economic benefit from bamboos is far beyond that of paddy (Zhu Zhaohua, 2007; from an unpublished article

Fig. 2.11a,b. *Guadua angustifolia* propagated from branches in Ecuador (Zhu Zhaohua).

Fig. 2.12a,b. Propagation of bamboo by seed, showing seeds and sprouted seedlings (Sun Maosheng).

used in the training course '*Evaluation on the Bamboo Industry's Impact on Rural Sustainable Development in Anji County*', p. 38). Ye Liangyou, a farmer in Anji County of Zhejiang Province, is good at managing both bamboo and paddy. He managed 0.4 ha of Moso bamboo forest and 0.2 ha of paddy fields. Tables 2.3 and 2.4 indicate that the net income of Moso bamboo forest was 30,155 CNY/ha every 2 years (an on year and an off year), i.e. 15,077.5 CNY/ha/year; while the net income from the paddy field was 5400 CNY/ha/year. So the net income generated from bamboo was 2.79 times more than that of paddy.

CASE STUDY 2: BAMBOO AS FUEL.

With its great potential, bamboo will become an important energy plant in the future.

A report by Growmore Bio-tech Ltd (Barathi, 2013) indicated that 64.8–81.0 ha of bamboo forest (*B. balcooa*) can provide materials to generate a million watts of electrical power a year. This capacity needs 400 ha of timber trees or the residues from more than 4000 ha of crop straw. Hence, bamboo has a great potential for being used as a renewable bioenergy plant.

Bamboo culms are the main resources for industrial processing

In 2010, China's yield of Moso bamboo culms was 1.4 billion poles, which was about 24 million tons. However, there is still a shortage in the supply of raw bamboo materials due to the rapid development of the bamboo sector. For example,

Fig. 2.13. Bamboo seedlings grown from seed (Sun Maosheng).

Fig. 2.14. Most bamboos were snapped off by a great snow in 2008 in Zhejiang, China (Lin Hai, President of Dasso, Zhejiang Province, China, 2014).

in 2010, Anji County's total Moso bamboo forest area was 52,000 ha and the total production of Moso culms was 25 million poles, but the actual total consumption was 130 million poles. Thus, 82% of the raw materials were imported from other provinces and counties nearby. The price of raw materials has been increasing year by year.

Fig. 2.15. The recovery within a year of bamboos snapped off by a great snow in 2008 in Zhejiang, China (Fig. 2.14 shows the snow damage caused) (Lin Hai, President of Dasso, Zhejiang Province, China, 2014).

The value increase of Moso bamboo culms in Anji County has indicated that bamboo forests and bamboo shoot production can create huge wealth for the people. In 2010, Anji's total output value for the bamboo sector was 13 billion CNY (US$1.985 billion). In the same year, China's total production of bamboo panels was 3.58 million tons, of which there were 1.11 million m³ of bamboo floors, 3.37 million tons of daily goods, 2.17 million tons of bamboo pulp, 11.25 million pieces of bamboo furniture and 0.12 million tons bamboo fibre products. The total value of China's bamboo sector was 117.3 billion CNY (US$18.2 billion) in 2010, and this increased to 167.1 billion CNY (US$26.95 billion) in 2013. It is predicted that the total value of China's bamboo sector will reach 300 billion CNY by 2020, which will be 2.6 times more than its value in 2010 (data from the China State Forestry Administration, 2013).

To conclude, as an alternative to timber, bamboo poses huge potential in the various fields of engineered boards, paper and pulp, furniture and construction. The applications of laminated bamboo panels and pressed bamboo lumbers especially have made it possible for bamboo to widely replace timber. The technology of bamboo pressing has made it possible to press the bamboo materials into lumber that basically has a performance that is the same as (tree-produced) wood lumber, and can be processed into almost all the products that can be produced from lumber. By 2012, China had 150 production lines of pressed bamboo lumber, each with an annual production capacity of 4000 m³, and the total annual consumption of Moso bamboo culms reached 20,000 tons.

Bamboos provide a large amount of green foods

A farmer household in Lin'an County, Zhejiang Province created a highest record of 100,000 US$/ha a year for bamboo shoot production.

Most bamboo species produce edible shoots; and many of them are green foods, as they come from natural bamboo forests or plantations to which chemical fertilizers and pesticides are

Fig. 2.16. *Dendrocalamus giganteus* in Yunnan Province, China (Zhu Zhaohua).

never applied. Bamboo shoots contain vitamins, amino acids, cellulose and many trace elements that are good for the human body. Bamboo shoots are served as delicious vegetable dishes in many countries, especially China, Japan and a number of South-east Asian countries. Studies have shown that bamboo shoots are one of the ideal vegetables containing high protein (2–4%) and amino acids, and have a low fat content and high contents of dietary fibre and mineral elements. In particular, the bamboo shoots of some bamboo species, such as Moso bamboo, are rich in selenium (0.058–2.65 µg/g), which is one of the essential elements for life, but is lacking in many regions in the world. While many bamboo species produce edible shoots, the most well-known of these can be placed in four groups.

1. Many species in the genus *Phyllostachys* are fine species for shoot production purposes, including: *Ph. acuta, Ph. heterocycla* var. *pubescens, Ph. praecox, Ph. prominens, Ph. vivax, Ph. robustiramea; Ph. dulcis, Ph. elegans, Ph. glabrata, Ph. flexuosa,*

Ph. glauca, Ph. aurea (a monopodial bamboo that can grow in tropical areas) and *Ph. nuda* (good for making *Tian Mu,* or dried bamboo shoot).
2. Sympodial species in the genus *Dendrocalamus* are also suitable for shoot production, including: *D. latiflorus, D. asper, D. brandisii, D. variostriata, D. giganteus, D. membranaceus* and *D. semiscandens.*
3. Other important edible shoot species include: *Fargesia canaliculata* (a species growing in high mountains that both giant pandas and humans love to eat); *Dendrocalamopsis oldhami* (also known as horseshoe-shaped bamboo shoot), *Chimonobambusa quadrangularis* (the culm is square shaped), *Indosasa sinica* (a monopodial bamboo species grown in tropical mountainous regions), *Schizostachyum funghomi* (a very common wild tropical bamboo species) and *Melocalamus arrectus* (a climbing bamboo species.
4. Some tropical bamboo species belonging to the genus *Bambusa* also have edible shoots; the most well-known is *B. blumeana.* Recent studies in Chile showed that *Chusquea culeou,* which is grown in Latin America and has a solid culm (see Fig. 2.28), also produces edible shoots (Zhu Shilin and Ma Naixun, 1993).

The bamboo shoot industry in China has been developing as fast as the bamboo culm processing industry. In 2010, China produced around 5.5 million tons of fresh bamboo shoots in total; the majority (about 70%) of them was sold directly in local and domestic markets, but a few fresh shoots were exported to Japan, South Korea and South-east Asian countries. About 30% of the fresh bamboo shoots were processed into preserved products, such as boiled shoots, canned shoots, dried shoots, fermented shoots and instant shoots. Some 40% of preserved bamboo shoots were exported to the international market and the remaining 60% were sold domestically. In 2011, China's total production of preserved bamboo shoots was 1.66 million tons (data from the China State Forestry Administration, 2013).

Lin'an County in Zhejiang Province has the largest production of preserved bamboo shoots in China. The shoot-oriented plantations in this county cover 35,000 ha, and in 2011, Lin'an's total production of fresh bamboo shoots was 220,000 tons, the total value of which reached 920 million CNY; including the value

Fig. 2.17. *Phyllostachys praecox* in Zhejiang Province, China (Zhu Zhaohua).

of processed shoot products, the total production of the bamboo shoot sector in Lin'an reached 2.2 billion CNY. In Lin'an, the dominant shoot-producing bamboo species are *Ph. vivax*, *Ph. praecox*, *Ph. prominens* and *Ph. nuda*. The average bamboo shoot forest areas owned per household in Lin'an were relatively small, less than 0.5 ha in most cases, with some even less than 0.1 ha. However, the economic benefits of bamboo shoot production are rather high. Statistics show that in Lin'an, the net income generated from bamboo shoots can reach 450,000 CNY/ha, sometimes up to 700,000 CNY/ha when applying covering techniques that can facilitate advanced shoot emergence from the ground. In 2010, there were 11,000 professional shoot-producing households in Lin'an whose income from fresh shoot sales exceeded 10,000 CNY, 4000 households with an income of more than 20,000 CNY, 180 households with an income exceeding 50,000 CNY, 120 households with an income more than 100,000 CNY, and three households with an income of even more than 200,000 CNY. Hence, the bamboo shoot industry has become

one of the pillar industries in Lin'an, which helps greatly in poverty alleviation (Zhu Zhaohua, 2009; from an unpublished article used in the training courses '*Multiple-participation in the Development of Bamboo Shoot Industry in Lin'an*').

In Tanzania, farmers have traditional techniques to make wine by directly using the juice of shoots of *Oxytenanthera abyssinica*. First, the farmers cut off the majority of bamboos from a clump, leaving a few stands that are of a certain distance from each other. When the new shoots up to 10–15 cm tall emerge, the farmers cut their tender tips to let the juice drain out, and using a bamboo sheath or the leaves, they guide the juice into a container specially made using bamboo internodes. To keep the juice flowing, the farmers need to cut the shoot tips three times a day – in the morning, noon and evening. Each shoot can generate 2 kg of juice every 24 h, and the juice can be collected from the same shoot for 30 days. During the 30 days, the bamboo shoots continue growing even though their tips are being cut. The juice can be directly used as a drink or salad sauce on the same day. On the second day, it will

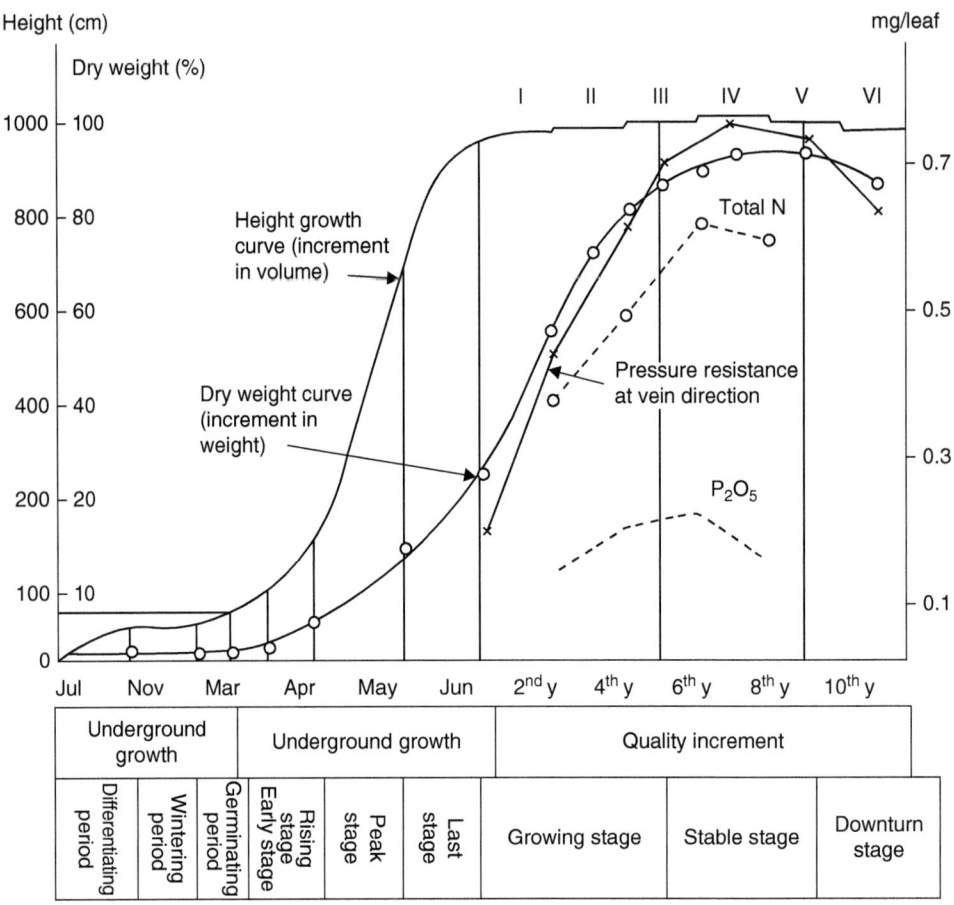

Fig. 2.18. Growing process of Moso bamboo (*Phyllostachys heterocycla* var. *pubescens*). P_2O_5 indicates phosphate content; 2nd y (etc.) indicates the 2nd year (etc.) (modified from Wu Danren, 1999, p. 84).

turn into a low-alcohol wine (similar to China's glutinous rice wine) by natural fermentation. The content of alcohol will increase with time of storage. The farmers can decide how long they would like to keep the bamboo shoot juice according to their own preferred taste.

In China, people have noticed the high dietary fibre in the bamboo shoots and have made puffed food such as high dietary fibre biscuits with other starch from other crops.

Various applications of bamboo leaves

China has a long history of using of bamboo leaves for food and medicinal purposes. The medical applications of bamboo leaves were carefully recorded in many pharmacopoeias in ancient times, such as Huang Gongxiu's *The Truth Seeking Materia Medica of China (Ben Cao Qiu*

Zhen, 《本草求真》) (AD 1769), which describes how bamboo leaves can cool fever, moderate the spleen, and reduce phlegm, coughing and a paralytic stroke. Bamboo leaves are among the many parts of bamboos with medicinal properties that are listed in the *Pharmacopoeia of the People's Republic of China* (or *Chinese Pharmacopoeia*). Bamboo sap, a natural liquid from heated fresh bamboo culms, and yellow chips (yellow parts on the inner layer of bamboo culm walls) are important traditional medicinal materials as well. For example, Professor Zhu experienced a heart surgery in 2006, and got a severe cough as a consequence. Western medicine and antibiotics did not help to stop the cough, but it was finally controlled by using bamboo sap.

China's Ministry of Health has listed bamboo leaves as a natural plant material for both food and medicinal uses. Bamboo leaves are also

Fig. 2.19. A 2-year-old bamboo energy forest in India, which can generate 100 tons biomass/ha annually (Barathi, 2013, with permission).

Fig. 2.20. An 8-month-old bamboo energy forest in India (from Barathi, 2013, with permission).

good fodder and make feed for animals, such as cattle and poultry. It is well known that bamboo leaves are the main food for giant pandas. In China, there are about 47 bamboo species in ten genera that can serve as good fodder for giant pandas, including monopodial species and sympodial species, and alpine species distributed in subtropical and temperate zones.

Studies on the active medicinal components of bamboo leaves reveal that bamboo

Table 2.1. Annual culm biomass varies with species (data from Ma Naixun, 2001).

Bamboo species	Annual culm biomass (t/ha)
Pseudosasa amabilis	15–20
Pleioblastus amarus	20
Phyllostachys heterocycla var. pubescens	30–40
Bambusa pervariabilis	40–45
Dendrocalamus membranaceus	40–50
B. rigida	45–50
B. chungii	60–80
Dendrocalamus giganteus	80–100

leaves are rich in flavonoids and other active ingredients. The amount of flavonoids in bamboo leaves can compare with that of *Ginkgo biloba*, from which a famous lipid-lowering drug is produced. Extracts of bamboo leaves have good effects against free radicals, as an antioxidant and anti-ageing agent, and in lowering blood lipid and reducing cholesterol. Bamboo leaves have great potential and value as foods (e.g. bamboo leaf rice), beverages (beers with flavonoids), bamboo leaf tea, medicine and, as already mentioned, as fodder and feed (Zhang Ying and Tong Lili, 1997).

Anji County ranks as one of the top counties in the utilization of bamboo leaves, and more than 70% of the bamboo leaves and branches produced are collected and fully used. To prevent damage to bamboo by snow, every autumn the farmers in Anji cut off the tips of the new culms of that year, and the cut-off top culm and branches and leaves are collected. The branches – still with leaves on – are then put into a simple facility to defoliate them. Anji's total bamboo leaves production can reach 30,000 tons/year, and they are used for three main purposes:

Fig. 2.21. Culm-oriented Moso bamboo (*Phyllostachys heterocycla* var. *pubescens*) forest (Zhu Zhaohua).

Fig. 2.22. Culm- and shoot-oriented bamboo forest (Zhu Zhaohua).

1. To cover the ground of *Ph. praecox* plantations in order to facilitate early shooting; for this, each hectare of plantation needs about 6 tons of bamboo leaves;
2. To be sent to biology technology companies, who will extract flavonoids for medicinal uses; and
3. To export as poultry feeds.

In 2005, the price of bamboo leaves in Anji reached 500–600 CNY/ton. Currently, bamboo leaves are in short supply for the market, and prices have been increasing continuously.

Uses of bamboo branches

The bamboo branches are usually left behind in the forest after harvesting, and need to be cleared up before they encourage the occurrence of pests and diseases and hamper management activities. Before the 1980s, most of the branches were cleared up by the farmers themselves, but since then, the branches have been fully utilized. According to the statistics for 2005, Anji produced 80,000 tons of branches, and the price of these was 700–850 CNY/ton. Among the total 80,000 tons, 50,000 tons were bamboo tips removed from new stands to prevent snow damage, and 30,000 tons were old branches collected from harvested bamboos. Bamboo branches are ideal materials for making brooms and besoms. Anji County has 200 bamboo branch processing factories, which produce over 8 million bamboo brooms and besoms every year, with an output value reaching 190 million CNY (US$30.6 million).

Uses of bamboo rhizomes, roots and culm bases

After harvesting, the base parts of the bamboo (generally 30–40 cm long) are usually cut off as waste materials, because they have highly condensed internodes, which makes them unsuitable for processing into floorings or split for weaving products. In the past, they could only be used for firewood, for which they have advantages such as thick walls and high density. Anji

Fig. 2.23. Shoot-oriented bamboo forest (Wang Kuihong, 2011).

Table 2.2. Characteristics of Moso bamboo (*Phyllostachys heterocycla* var. *pubescens*) forests raised for culms and/or shoots (data from Zhu Zhaohua, 2013).

Main targets	Culm production	Culm and shoot production	Shoot production
Density	3000–4000/ha	2300–2500/ha	1500–1800/ha
Management intensiveness	Soil loosening and weeding, once every 2 years; fertilizer, once a year	Soil loosening and weeding, once a year; fertilizer, twice a year; irrigation	Soil loosening and weeding, twice a year; fertilizer, three times; irrigation system installed
Income (e.g. in Anji)	18,000–23,000 CNY/2 years/ha (US$2951–3770)	52,500–75,000 CNY/2 years/ha (US$8607–12,295)	80,000–120,000 CNY/2 years/ha (US$13,114–19,672)
Harvest age	Above 5 years	Below 5 years	Below 4 years
Invested cost as % income	20–25%	30–35%	40–45%

produces 30,000 tons of such raw materials annually. The weight of the base part of a bamboo culm usually is about 8–9% of the total weight. Nowadays, the base parts are mainly purchased by charcoal-producing companies at the price of 200–300 CNY/ton. The base part also includes an underground part, which is connected to the rhizome by a special part called the rhizome neck (in Chinese, this is called 'the screw'); after the bamboo culm is harvested, this underground part has to be dug out to leave more space for rhizomes and new shoots to grow. The dug-out part is good material for handicraft making, such as bamboo carving craft. In China

Table 2.3. Analysis of input and output of 0.4 ha bamboo forest managed by Ye Liangyou in 2001–2002 (Zhu Zhaohua, 2007; see text for details).

Input	Input (CNY/0.4 ha)	Input (CNY/ha)	Product(s)	Income (CNY/0.4 ha)	Income/ha
Working days (per 2 years)	148	370	Culm	7400	18,500
Salary (CNY)	4440	11,100	Branches, tops and sheaths	1000	2500
Fertilizer (CNY)	1400	3500	Winter shoots	4390	10,975
			Spring shoots	2390	5975
			Rhizomes	2870	7175
Total (CNY)	5840	14,970	Total	18,050	45,125
Net income (CNY) = total revenue – total expenditure CNY/2 years/ha				12,210	30,155

Table 2.4. Analysis of input and output of 0.2 ha of paddy in a field managed by Ye Liangyou in 2001–2002 (Zhu Zhaohua, 2007; see text for details).

Input/output	Input (CNY/2 ha)	Input as (CNY)/ha	Output	Income (CNY/0.2 ha)	Income/ha
Soil	210	1050	Paddy	1800	9000
Seed	60	300	Straw	300	1500
Fertilizer	150	750			
Management	240	1200			
Pesticide	60	300			
Harvest	300	1500			
Total	1020	5100	Total	2100	10,500
Net income (CNY) = total revenue – total expenditure as CNY/ha/year				1080	5400

and some Latin American countries, such as Colombia, the underground base parts of bamboo are made into various types of exquisite art products.

The rhizomes of bamboos are also suitable for making arts and crafts. Anji has 11 factories specializing in rhizome processing; their products include handles, knobs for handbags and craft products. *Guadua angustifolia* (Guadua) is a widely distributed species in Latin America. It is an amphipodial species and the shape of its rhizome looks very much like a crocodile. Many craftworkers have found the Guadua rhizome to be a very good material for arts and crafts making. The fibrous roots growing on the rhizomes can also be made into handicrafts and utensils, such as scrubbing brushes (pot brushes); these are also collected and sold in large quantities in Anji.

It is worthy of mention that the above-mentioned parts are usually considered a waste in the traditional sense, but that they are now removed from the underground and processed into valuable arts and crafts. Not only the wastes are turned into treasure, but the removal of the old root parts of bamboo can help to loosen the soil, increase soil permeability, prevent the degeneration of the forest and promote the renewal of bamboos.

Uses of bamboo top parts and sheaths

TOP PARTS. Because of their small diameter and thin walls, the top parts of bamboos cannot be used for mat and panel processing. In Anji, every year, there are about 40,000 tons of bamboo top-part materials that are generated as waste from the primary processing factories. However, there are many factories that use the top parts for making scaffolding, chopsticks, toothpicks and fences. The market price for the top part of Moso bamboo was about 400–500 CNY in 2005.

Fig. 2.24. A bamboo biomass power plant in India (Barathi, 2013).

Fig. 2.25. Natural drying of *Guadua angustifolia* in Ecuador (Zhu Zhaohua).

Fig. 2.26. Bamboo loaded on a tractor in Lin'an, Zheijiang Province, China (Zhu Zhaohua).

Fig. 2.27. Canned *Chusquea* shoots with seafood (Chile) (Zhu Zhaohua).

Fig. 2.28. A natural *Chusquea culeou* bamboo forest in Chile (Zhu Zhaohua).

Fig. 2.29. An *Oxytenanthera abyssinica* plantation raised for wine in Tanzania (Zhu Zhaohua).

Fig. 2.30. In Tanzania, the production of bamboo juice is induced by cutting the shoot tops. After natural fermentation, it becomes wine (Zhu Zhaohua).

SHEATHS. Sheaths are the covers that wrap the new shoots of bamboos, and they fall off when the shoots grow into full-sized stands. The sheaths can be used for feeding to cows. In Lin'an, Zhejiang Province, the fresh sheaths of bamboo shoots are used for fodder for milking cows; they can be fed to the cows either fresh, or fermented and made into feeds. In early summer, when the sheaths fall from the new stands of bamboos, the bamboo farmers will collect them and sell them in the markets. They are mainly used for sofa cushions, snack boxes, pulp and handicraft processing and production. The value created by utilizing sheaths can reach to 1000 CNY/ha per every 2 years.

We can see that bamboo is a treasure, and that it has great potential for development. Its real economic value depends on the extent and degree of its innovative uses, the exploration of its potential and creativities in new product development; this process could go on and on infinitely.

A STRIKING EXAMPLE OF BAMBOO USE. A very striking example of bamboo use was in Sichuan Province, China. In the 1990s, a bamboo craft Master named Chen Yunhua from Qingshen County worked with six of his apprentices for more than a year, and used more than 200 million fine bamboo slivers to complete a masterpiece of plain weaving art – a 5 m long piece of woven art imitating the famous ancient painting of the Song Dynasty (AD 960–1279): *Along the River During Qingming Festival (Qing Ming Shang He Tu)*. In this woven art, there were more than 800 portraits of people, and more than 200 animals. Yet the total weight of this masterpiece was less than 300 g, and it was as thin as onion skin and smooth as silk, and as authentic as the original painting. A Taiwanese businessman bought this piece of bamboo woven art at the price of 1.06 million CNY (US$171,000). The raw material used for this masterpiece weighed not more than 100 kg, and came from a widely distributed local bamboo species – *Neosinocalamus affinis*, a medium-sized sympodial bamboo species. This type of species has features such as long internodes, even and smooth nodes and thin walls, and is easy to split. There are many bamboo species that have similar properties, and can be found almost everywhere in Asia. The key is to have certain capacities and skills to develop those bamboos.

2.3 Bamboos Are Important Members of the Ecosystem

2.3.1 Bamboo forest has strong capacity in water and soil conservation

Bamboos have strong capacity for adaptation, making them perfectly suitable for conserving small watersheds, hillsides with serious water and soil erosion, riverbanks and water reservoirs; they can be especially effective in holding in place riverbanks and hillsides. According to research done by Fujian Forestry College, in a Moso bamboo forest, for every cubic metre of soil below the ground, 38–46% is occupied by the bamboo root system (including its rhizomes and its diffuse and fibrous roots). Research has also shown that the soil-binding capacity of Moso bamboo forest is 1.5 times that of *Pinus massoniana*, its precipitation absorbing capacity is 1.3 times

Fig. 2.31. Roadside vendors of bamboo wine in Tanzania (Zhu Zhaohua).

that of Chinese fir (*Cunninghamia lanceolata*) and its water conservation capacity is 30–35% more than that of Chinese fir; overall, the conservation capacity of a Moso bamboo forest could be 1.5 times that of Chinese fir.

Other research has revealed that mixed forest of bamboo and other tree species performs better in integrated environmental protection than pure bamboo forest. A new bamboo forest is easy to establish and reaches maturity in 5 years, and with proper management, 25% of the forest can be harvested every year. Besides the culms, the bamboo forest also offers bamboo shoots as foods. A well-managed bamboo forest can provide economic benefits continuously, while also developing and providing sustainable eco-services. Bamboo has strong capacity for self-regeneration and can revive and prosper as a bamboo forest even after it has been clear-cut

or subject to forest fire. This is a significant difference from the trees, which after a clear-cut need at least several decades to recover their capacities in all aspects of eco-services and generate economic benefits. Bamboo can play an essential role in riverbank protection, shelter forest establishment and watershed management.

2.3.2 Bamboo and climate change

China's total moso bamboo area is only 2.78% of the total forest area of the country, but the bamboo forests of China absorb 11.6% of the carbon.

Fast growing with high biomass, bamboo has huge potential to absorb more carbon than other plants. In general, bamboo's capacities

Fig. 2.32. Bamboo leaves being removed from branches, Anji, Zhejiang Province, China (Zhu Zhaohua).

Fig. 2.33. Beverages containing bamboo flavonoids (Zhu Zhaohua).

Fig. 2.34. Bamboo branches ready for transportation to a factory in Anji, Zhejiang Province, China (Zhu Zhaohua).

Fig. 2.35. Bamboo branches being processed into brooms in Anji, Zhejiang Province, China (Zhu Zhaohua).

Fig. 2.36. Rhizome distribution in a *Guadua angustifolia* plantation in Colombia (Zhu Zhaohua).

Fig. 2.37. Beautiful Guadua bamboo rhizomes in Colombia (Zhu Zhaohua).

of carbon dioxide sequestration and oxygen release are 35% higher than those of trees. Moreover, carbon stored in bamboo can last for several decades in the form of durable bamboo products with high value, such as bamboo housing, bamboo furniture, bamboo panels, bamboo fibres, mats, handicrafts, etc. Hence, the development of bamboo can contribute to preventing the greenhouse effect and fighting against climate change (Joshi, 2013).

Take China for instance. China has placed great importance on bamboo sector development. China's bamboo forest area has spread from 2.87 million ha in the 1950s to 5.38 million ha in 2010, of which 3.87 million ha are Moso bamboo forest. Bamboo serves the equivalent of 15% of China's total tree resources, but its area only accounts for 2.78% of the total forest area. According to systematic studies by Zhou Guomo and Shi Yongjun

Fig. 2.38. Guadua rhizome handicrafts in Costa Rica (I.V. Ramanuja Rao).

Fig. 2.39. Guadua rhizome handicrafts in Ecuador (Zhu Zhaohua).

(2012), the total carbon storage of bamboo forests increased from 286.60 Tg C in the 1950s to 672.70 Tg C in 2008, which accounts for 11.6% of total forest carbon storage. Although bamboo forest only accounts for 2.78% of the total forest area, the annual increase in bamboo forest area has risen from 50,000 ha in the 1980s to 100,000 ha after 2010. Research also revealed that bamboo can sequestrate more carbon when it is managed by scientific harvesting and shoot promotion. Moso bamboo forest under regular harvesting accumulated 1.34 times more carbon than Chinese fir (the main timber species in China, with rotation of 20–30 years for harvesting) and 1.87 times that of *Pinus massoniana* at 60 years old. The capacity of carbon sequestration of *D. latiflorus* is equivalent to that of fast-growing eucalyptus. Of the carbon stored in a bamboo forest of which an average of 20–25% of the total bamboo stands are harvested annually, an average of 37% of carbon is transferred from raw bamboo into bamboo products.

Fig. 2.40. Moso bamboo (*Phyllostachys heterocycla* var. *pubescens*) sheath handicrafts (Jin Wei).

2.3.3 A more stable management model for the bamboo forest ecosystem

In most cases around the world, bamboo grows in natural forest mixed with other tree species. If managed properly (by timely and rational harvesting), bamboo can keep a high productivity when it is adapted to the site conditions, even though it is under extensive management. Moreover, very few pests and diseases occur, so the whole ecosystem can be balanced by the bamboo and other trees. In the development of plantations, to create a more comfortable environment for growth, improve the soil and increase famers' incomes, intercropping of bamboo with other crops is usually done when the plantations

are still new and young. The crops can be beans, peanuts, potatoes, medicinal plants, ornamental plants and so on. Usually, intercropping is done from years 1–5 after planting Moso and other subtropical bamboo species, while tropical bamboo species are generally intercropped in the first 1–3 years, and bamboo forests raised for shoot production are intercropped for the first 2–3 years. In China, since the 1980s, as people paid more attention to the development of bamboo forests, high investments have been made to establish large areas of monocultured, high-yielding bamboo forests, with high density and intensive management. The results were a decline in bamboo forest productivity and frequent pest and disease attacks. The lessons are that attention

Fig. 2.41. Utensils in daily use made from sheaths of Moso bamboo (*Phyllostachys heterocycla* var. *pubescens*) (Zhu Zhaohua).

Fig. 2.42. *Bambusa textilis* for river bank conservation, Guangning, Guangdong Province, China (Photo by Forestry Bureau of Guangning County, Guangdong Province, China).

Fig. 2.43. Distribution of Moso bamboo (*Phyllostachys heterocycla* var. *pubescens*) rhizomes and roots (Lin Hai, President of Dasso, Zhejiang Province, China).

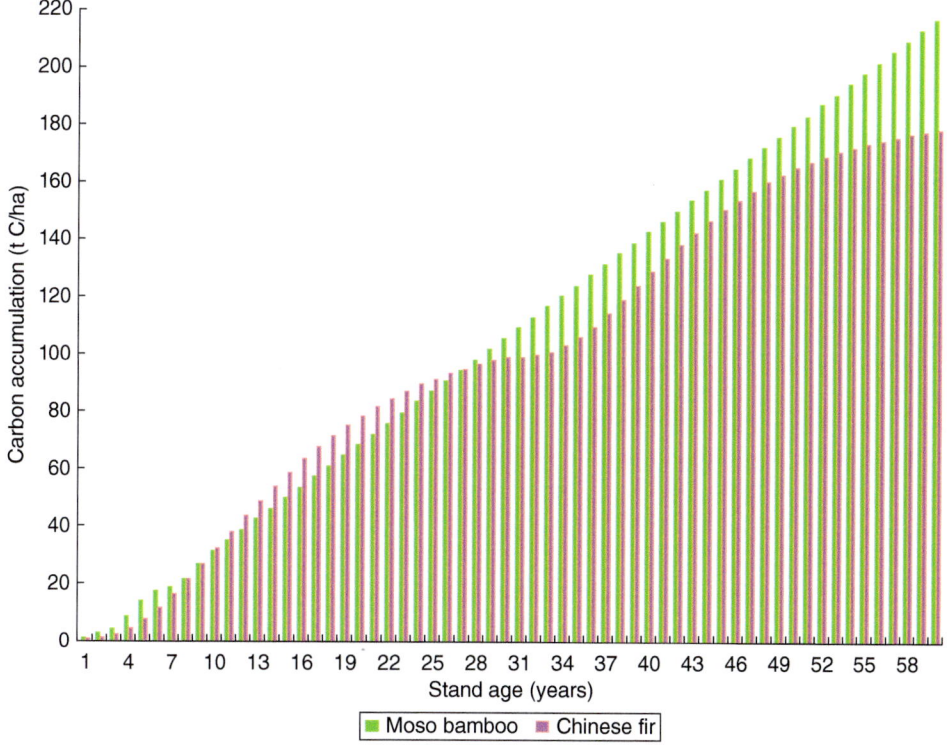

Fig. 2.44. Comparison of carbon accumulation by managed stands of Moso bamboo (*Phyllostachys heterocycla* var. *pubescens*) and Chinese fir (*Cunninghamia lanceolata*) over 60 years (Li Yanxia, INBAR).

Fig. 2.45. Natural bamboo and timber tree mixed forest, Laos (Zhu Zhaohua).

should be paid to the sustainability of the ecosystems in bamboo forests, and that mixed forests of bamboo and other timber trees should be promoted.

2.4 Bamboo Materials Have Certain Advantages Over Tree Timbers, which They Can Widely Replace and So Reduce Deforestation

2.4.1 Structure and main characteristics of bamboo materials

People often say that bamboo can replace wood from timber trees, and sometimes bamboo has certain advantages over tree timber, and can even be made into products that cannot be

made from wood from trees. However, bamboo has a number of problems while being processed, including a high vulnerability to insect and mould attacks, its small diameter and that it is easy to crack. These problems need to be solved through comparatively more complicated and professional treatment and processing. The above-mentioned advantages and defects are determined by the natural characteristics and the structures of bamboo materials.

Morphological characteristics of bamboos

The form of bamboos and trees are quite different. For example, the bamboo culm, which is the above-ground part of the stem, is composed of three parts – the stem petiole, stem base and stem proper. The rhizome neck (rural China calls it

Fig. 2.46. Moso bamboo (*Phyllostachys heterocycla* var. *pubescens*) and broadleaved tree (*Sassafras tzumu*) mixed plantation in Zhejiang Province, China (Zhu Zhaohua).

Fig. 2.47. Intercropping of tea and bamboo (*Phyllostachys vivax*) in Lin'an, Zhejiang Province, China (Zhu Zhaohua).

'the screw') connects the mother stock and the rhizomes. The stem base is the underground part of the culm on which the rhizome and roots grow; this part is composed of more than a dozen very short internodes. The rhizome neck, culm base and root parts together are called the bamboo root ball. The stems are above ground, are usually straight and hollow and have internodes. In the above part of a culm, there are branches and leaves. Usually every internode has two axial branches, the lower sheath axial branches and the upper sheath axial branches. Inside the two axial branches is the inner part of the node, and in between the axial branches are the internodes.

Table 2.5 shows the differences in growth characteristics between bamboos and timber trees.

Bamboo material structure

The term 'bamboo raw materials' mainly refers to the bamboo culms, which are the parts of the bamboo with the highest value. The majority of bamboo culms are hollow inside, and the hollow is usually called the medullary cavity, but there are a number of bamboo species with solid or nearly solid culms. *Chusquea*, which is distributed in Latin America, is one of these solid-culmed bamboo species. The culm cavity, where present, is surrounded by bamboo walls, the walls can be divided into three layers (for processing purposes): the green parts, pulp parts and yellow parts. Bamboos have no cambium. The culm does not grow wider, there is no pith and there are no radial cells, all cells are arranged in parallel with the wall direction; the vascular bundles are distributed inside the tissues, so the culm is easy to split. This is a very different structure from tree wood. That is why the bamboo, unlike tree wood, can very easily be split into strips, and be made into different products according to the properties and structures of the different layers of the bamboo material.

The physical properties of bamboos

BAMBOO DENSITY. In China, the density of bamboo (as g/cm^2) is measured using basic volumetric weight values as the ratio between the absolute dry weight and the maximum volume of the bamboo after it is soaked. The formula used is:

$$\frac{G_3 - (G_2 - G_1)}{V_{max}}$$

where G_1 = weight (g) before drying and volume measurement, G_2 = weight (g) before drying and after volume measurement, G_3 = dry weight (g) and V_{max} = maximum volume (cm^3) after soaking. Tables 2.6 and 2.7 present data on the density and expansion in water of different bamboo species. The data in Table 2.7 show that bamboo density increases with age from 1 to 6 years old, and that the density at ages >6 years reduces (Wu Danren, 1999, Table 5.1.2, p. 294).

WATER ABSORPTION BY BAMBOO. The absorption and evaporation of water by bamboo are two

Table 2.5. Differences between bamboos and timber trees (from Zhang Qisheng* et al., 2002, p. 7).

Growth parameter	Bamboo	Timber tree
Height growth	1. Height growth completes within 2–4 months. 2. It is mainly by means of intercalary meristems. 3. Height growth begins and ends in different internodes non-simultaneously, and the whole growth process is realized by intercalary meristems.	1. Height growth lasts the whole lifetime of the tree, and the speed of growth declines with ageing. 2. Height growth is realized by the primary meristem on the apex. 3. Height growth does not take place in secondary growth tissue.
Diameter growth	1. In the process of height growth of the shoot into the young bamboo, the diameter of the stem and the thickness of the stem wall increases slightly. 2. After the completion of height growth, the diameter does not increase.	1. Diameter growth is realized by the cambium. 2. It lasts the whole lifetime of the tree.

*Professor Zhang Qisheng unfortunately passed away in September 2017.

Table 2.6. Density, moisture content and water expansion ratios of different bamboo materials (data from Wu Danren, 1999).

Species	Density (g/cm³)	Moisture (%)	Expansion (%)			
			Tangential direction	Radial direction	Vertical direction	Volume
Dendrocalamus bicicatricata	0.339	193.14	4.53	8.45	0.30	13.71
D. latiflorus	0.372	182.33	6.93	12.53	0.23	20.60
Bambusa chungii	0.418	186.31	8.80	24.84	0.52	36.54
B. oldhamii	0.443	134.68	11.64	16.68	0.31	28.37
B. flexuosa	0.454	106.80	6.27	7.02	0.35	14.27
B. pervariabilis	0.456	135.93	7.81	11.78	0.16	20.69
B. tulda	0.505	103.45	6.98	6.83	0.29	14.63
Phyllostachys (Ph.) bambusoides	0.508	107.24	6.36	6.81	0.34	13.98
Lingnania fimbriligulata	0.551	96.68	6.16	11.82	0.15	17.65
B. rigida	0.553	87.17	7.19	9.24	0.15	17.25
Ph. heteroclada	0.572	83.02	5.15	4.79	0.26	10.51
B. vulgaris cv. Vittata	0.574	73.51	4.34	5.03	0.24	11.59
Indosasa crassiflora	0.590	71.14	2.56	5.88	0.14	8.98
B. vulgaris	0.591	70.45	5.51	4.70	0.42	10.84
Ph. pubescens	0.614	83.31	5.77	9.87	0.00	19.31
B. textilis	0.629	79.89	5.57	17.05	0.22	23.82
Ph. glauca	0.632	58.90	3.13	4.10	0.48	7.82
Ph. flexuosa	0.656	74.50	3.16	13.19	0.17	17.02
Ph. glauca var. *variabilis*	0.656	60.76	4.24	4.73	1.45	9.55
Ph. viridis	0.660	53.36	3.73	6.49	0.29	10.76
Pseudosasa amabilis	0.675	56.19	4.24	7.49	0.24	11.52
Ph. bambusoides	0.679	59.24	4.85	5.26	0.27	10.59
Ph. meyeri	0.721	55.97	4.01	5.76	0.393	10.67

Table 2.7. The density of Moso bamboo (*Phyllostachys heterocycla* var. *pubescens*) at different ages (data from Wu Danren, 1999, Table 5.1.2, p. 294).

Age (years)	1–2	3–4	5–6	7–8	9–10
Density (g/cm³)	0.529	0.696	0.627	0.626	0.600

opposite processes. Dry bamboo has a strong water absorption capacity, and the water absorption rate is inversely proportional to the length of the bamboo and has little to do with its width. This phenomenon indicates that the absorption and evaporation of water by bamboo is primarily conducted through its cross section. The size and volume of bamboo, just like that of (tree) wood, will expand in all directions after the absorption of water, and the strength will then be relatively reduced. Before the material reaches its fibre saturation point, its strength and moisture content increase, while after it has reached the saturation point, the water content continues to increase, but the strength changes little, thus causing a relative reduction in strength. When bamboo is in an absolutely dry state with a brittle texture, its strength decreases. Table 2.8 gives data on the shrinkage of (dried) bamboo and shows the resulting decrease in strength (density).

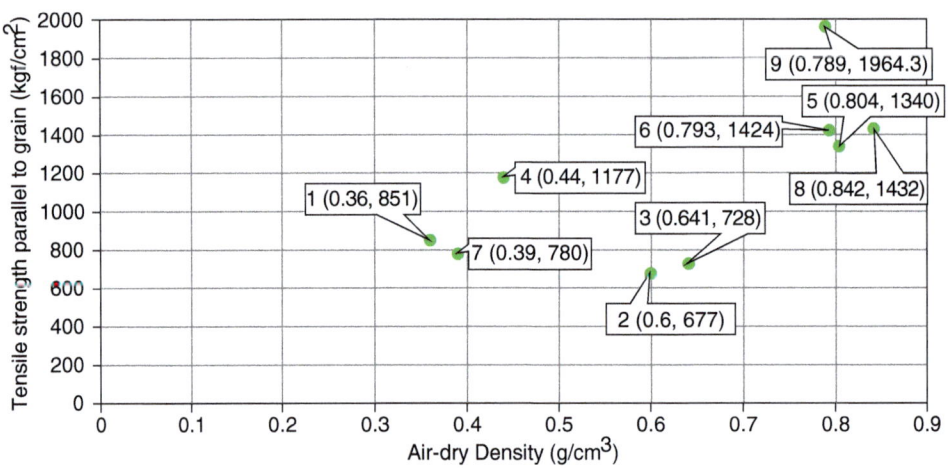

Fig. 2.48. From the chart above, it could be indicated that the tensile strength parallel to grain of moso bamboo is distinguishably higher than the other major timber tree species. (The source of the figures in this chart is *Wood Physical and Mechanical Properties of Main Tree Species In China* (中国主要树种木材物理力学性质), edited by Wood Research Institute of the China Forestry Academy, published by China Forestry Publishing House in 1982). Key: 1. *Cunninghamia lanccolate* (杉木); 2. *Cupressus duclouxiana* (柏木); 3. *Larix gmellini* (落叶松); 4. *Pinus koraiensis* (红松); 5. *Eucalyptus camaldulensis* (赤桉); 6. *Fagus longipetiolata* (水青冈); 7. *Populus ussuriensis* (大青杨); 8. *Querus acutissima* (麻栎); 9. *Phyllostachys heterocycla var. pubescens* (moso bamboo, 毛竹).

Table 2.8. Shrinkage of Moso bamboo (*Phyllostachys heterocycla* var. *pubescens*) and *Neosinocalamus affinis* on drying (data from Wu Danren, 1999, p. 295).

	Shrinkage (%)					
Species	Radial	Tangential	Vertical	Volume	Outer circumference	Inner circumference
Moso bamboo	4 ± 0.5	7.5 ± 0.5	0.15	1.0	–	–
N. affinis	4 ± 0.5	3 ± 0.5	–	1 ± 0.5	5 ± 0.5	5 ± 0.5

MECHANICAL PROPERTIES OF BAMBOO. Bamboo usually has good rigidity, so they are naturally good engineering materials for construction. As it is also easy to split, processing can be done both by hand and by machine. For thousands of years, bamboos were utilized to produce many daily necessities and handicrafts that could not be made from other natural materials. The engineering strength properties of bamboo are large; the resistance strength is about two times that of (tree) wood, and the compression strength is about 10% higher than that of (tree) wood. Bamboo is an anisotropic material, and when the compression direction is different; the compression strength is also different. The compression strength of bamboo is a vital feature for its use as a structural material, yet various factors can affect the stability of the mechanical strength of bamboo, such as species, standing conditions, age of the culm and part of the culm. Table 2.9 indicates how species affects the mechanical strength of bamboo and (tree) wood and compares these values with those of various types of steel.

The mechanical strength of bamboo material varies as a result of numerous factors, including species, site (climatic) conditions and bamboo age, as shown in Tables 2.10–2.13.

Chemical properties of bamboos

The main chemical components of bamboo are cellulose, hemicellulose and lignin, and there are also a variety of carbohydrates, fats and protein substances. The chemical composition of various species of bamboo is summarized in Tables 2.14 and 2.15.

CELLULOSE CONTENT (AND THE ASSOCIATED FIBRE PROPERTIES).

> *Bamboos can be an important source for future natural fibre.*

The cellulose contents of bamboos are usually 40–60% (see Table 2.15). Different species and different aged bamboos may have different cellulose contents. For example, the cellulose content of the culm of young Moso bamboo (a subtropical species) at <1 year old is 75%, at 1 year old it is 66% and at 3 years old it is 58%; in contrast,

Table 2.9. Comparison of strength of bamboo, (tree) wood and steel (data from Wu Danren, 1999).

		Tensile strength (kg/cm²)		Compression strength (kg/cm²)	
Species		By species	Averages for group	By species	Averages for group
Bamboo species	Moso bamboo (*Phyllostachys heterocycla* var. *pubescens*)	1984.2	2082.2	640.0	487.2
	Ph. viridis	2833.5		540.0	
	Ph. meyeri	1821.8		359.6	
	Dendrocalamus latiflorus	1951.2		411.3	
Timber species	Chinese fir (*Cunninghamia lanceolata*)	712	1007.7	406	428.5
	Chestnut (*Fagus*)	984		358	
	Pine (*Pinus*)	1300		525	
	Birch (*Betula*)	1035		426	
Steel	Mild steel	3784–4250	5170–5563 or above	As in the previous column (the tensile strengths of metals are the same as their compression strengths)	
	Semi-mild steel	4400–5000			
	Semi-hard steel	5200–6000			
	Hard steel	7300 or above			

Table 2.10. Physical properties of several bamboo species (data from Zhang Qisheng, 1995).

Physical properties	Moso bamboo (*Phyllostachys heterocycla* var. *pubescens*)	*Neosinocalamus affinis*	*Dendrocalamus asper*	*Phyllostachys glauca*	*Phyllostachys iridis*
Tensile strength (MPa)	187.77	227.55	199.10	185.89	289.13
Static bending strength (MPa)	163.90	–	–	213.36	194.08

Table 2.11. Impacts of climatic conditions on the physical properties of Moso bamboo (*Phyllostachys heterocycla* var. *pubescens*) (data from Zhang Qisheng, 1995).

Location	Longitude E.	Latitude N.	Annual average temperature (°C)	Annual average precipitation (mm)	Tensile strength (MPa)	Compression strength (Mpa)
Yixing[a]	119°51′	30°0′	15.60	1320	200.06	71.96
Shimen[b]	121°16′	29°37′	15.98	1512	185.67	61.17
Damaoshan[c]	117°48′	28°45′	17.60	1800	177.54	61.15

[a]Jiangsu Province, China; [b]Zhejiang Province, China; [c]Jiangxi Province, China.

Table 2.12. Impact of age on physical properties of Moso bamboo (*Phyllostachys heterocycla* var. *pubescens*), Jiangsu Province, China (data from Zhang Qisheng, 1995).

Physical strength property	Age (years)										
	<1	1	2	3	4	5	6	7	8	9	10
Tensile strength (MPa)	–	135.35	174.76	195.55	186.15	184.83	180.64	192.40	214.93	185.70	185.61
Compression strength (MPa)	18.48	49.05	60.61	65.38	69.51	67.53	69.51	67.45	75.51	64.89	62.68

Table 2.13. The relationship between height on the culm and the physical properties of Moso bamboo (*Phyllostachys heterocycla* var. *pubescens*) produced in Puqi and Chongyang, Hubei Province (data from Zhang Qisheng, 1995).

Strength property	With/ without node(s)	Height on culm (m)						
		1	2	3	4	5	6	7
Tensile strength (MPa)	With	126.84	147.73	167.34	166.94	167.55	169.90	169.49
	Without	157.96	191.02	194.28	202.14	208.98	215.41	221.22
Static bending strength (MPa)	With	140.31	149.79	151.84	156.12	162.86	173.26	172.45
	Without	138.77	147.35	152.14	152.75	160.82	162.04	170.20

Table 2.14. Extract contents of different bamboo species (data from Wu Danren, 1999).

Extract content (%)	Moso bamboo (*Phyllostachys heterocycla* var. *pubescens*)	*Phyllostachys glauca*	*Bambusa pervariabilis*	*Neosinocalamus affinis*	*Dendrocalamus asper*
Cold water extract	2.60	–	4.29	–	–
Hot water extract	5.65	7.65	5.30	–	12.41
Alcohol and ethyl ether extract	3.67	–	5.44	–	–
Alcohol benzene extraction	–	5.74	3.55	8.91	6.66
1%NaOH extract	30.98	29.95	29.12	27.62	21.81

Table 2.15. Range of contents of chemical and extracts found in bamboos (data from Zhang Qisheng, 1995).

Content (%)	Chemical/extract					
	Cellulose	Pentosan	Lignin	Cold water extract	Hot water extract	Alcohol and ethyl ether extract
Range	(40–60)	(14–25)	(16–34)	(2.5–5.0)	(5.0–12.5)	(3.5–5.5)
Average	50.38	20.86	25.45	3.92	7.72	4.55

Content (%)	Chemical/extract					
	Alcohol benzene extract	1% NaOH extract	Protein	Fat and wax	Starch	Reducing sugars
Range	(2–9)	(21–31)	(1.5–6)	(2–4)	(2–6)	(~2.0)
Average	5.45	27.26	2.55	2.87	3.60	2.0

Content (%)	Chemical/extract					
	Nitrogen	P_2O_5	K_2O	SiO_2	Other ash elements	Total ash
Range	(0.21–0.26)	(0.11–0.24)	(0.5–1.2)	(0.1–0.5)	(0.3–1.3)	(1.0~3.5)
Average	0.24	0.16	0.82	1.30	0.72	2.04

Fig. 2.49. 100% bamboo pulp tissue manufacturing in Qingshen, Sichuan Province, China (Zhu Zhaohua).

for the tropical species *Dendrocalamus latiflorus*, the cellulose content of the culm at 1 year old is 53.19%, at 2 years old it is 52.78% and at 3 years old it is 50.77%. Usually, as the bamboos grow older, their cellulose contents decrease. A famous Chinese bamboo researcher, Ma Naixun thoroughly researched the cellulose content of different bamboo species (Ma Naixun, 2001,

p. 7, Table 6). The quality of bamboo cellulose is quite good, and the resource is rich and suitable for large-scale natural fibre product development. There are large-scale bamboo pulp and paper industries in India, China, Vietnam and Brazil, etc. Table 2.16 shows that different bamboo species have different fibre lengths, cellulose contents and fibre length-to-width ratios. Prof. Ma Naixun selected 33 fine bamboo species for pulp making from 102 candidate species selected from China; of these, 31 species were sympodial and only two were monopodial. There were 21 best species, and these included B. papillata, B. textilis, N. affinis, N. affinis cv. Flavidorivens and Dendrocalamopsis minor var. amoenus; these were followed by species such as: Dendrocalamopsis daii, Ph. meyeri, Pleioblastus gozadakensis and B. ventricosa (Ma Naixun, 2001).

From the late 1990s, China began extensive use of fine bamboo species for clothing, decorative and industrial fibre products. In 2010, its bamboo pulp production reached 2.17 million tons and bamboo fibre products reached 120,000 tons. It takes about 3 tons of bamboos to produce 1 ton of natural fibre. The bamboo pulp and fibre industry has shown broad prospects.

Brazil is very successful among the Latin American countries in growing bamboo pulp oriented plantations. Since the 1970s, Brazil has planted four large paper pulp oriented bamboo forests, mainly in northern Brazil (in the states of Maranhão, Piauí, Paraíba and Pernambuco). Up to 2014, the total area of bamboo pulp oriented plantations reached 40,000 ha. The main species planted is B. vulgaris cv. Vittata, which has advantages such as being easy to propagate, growing faster after it has been cut, flowers rarely, has adequate productivity, a wide adaption to the climatic and soil conditions of Brazil, and has good fibre features, etc. Clear-cutting of the plantations occurs every 2 years (for a new plantation this is 3 years), and the clear-cutting output reaches 60 tons/ha on average. The AGRIMEX Company clear-cuts 20 ha a day on average, which provides 18,000 tons of papermaking raw materials monthly. After

Table 2.16. A comparison of the fibre characteristics and chemical composition of different bamboo, tree and herbaceous species used in pulp making (data from Ma Naixun, 2001, p. 54).

Species	Fibre length (mm)	Fibre length to width ratio	Fibre chemical composition (%)		
			Cellulose	Lignocellulose	Ash
Bamboos					
Neosinocalamus affinis	2.71	198.8	44.35	31.28	1.20
Bambusa rigida	2.06	177.6	49.96	22.83	2.91
B. pervariabilis	2.34	196.1	49.79	24.83	4.53
B. textilis	3.04	221.4	52.16	24.03	2.16
B. chungii	2.90	256.0	48.76	27.73	4.06
Dendrocalamopsis oldhami	2.48	180.1	49.55	23.00	1.78
Schizostachyum funghomii	3.53	296.0	–	–	–
Phyllostachys pubescens	2.25	165.1	45.50	30.67	1.10
Pleioblastus amarus	2.18	144.3	44.55	25.33	1.51
Trees					
Pinus massoniana	3.61	72	51.86	28.42	0.33
Larix spp.	3.41	77	52.55	27.44	0.36
Pinus elliottii	2.93	73	76.78[a]	28.34	
Populus tomentosa	0.82	39.5	78.85[a]	23.75	0.84
Eucalyptus citriodora	0.94	64	77.80[a]	27.45	0.29
Herbaceous species					
Phragmites communis	1.12	115	43.55	25.40	2.96
Oryza sativa	0.92	114	36.20	14.05	5.50
Triticum aestivum	1.32	102	40.40	22.34	6.04

[a]Content of compound cellulose.

every clear-cutting, the land is fertilized to facilitate the regrowth of the bamboo forests. With this technology, the productivity of this company's bamboo plantations has never been decreased over 40 years, except in drought or when insufficient fertilizer was applied.

HEMICELLULOSE CONTENT. The hemicellulose content of bamboos is about 14–25%, and varies with species and age. For example, the hemicellulose content of Moso bamboo is around 22.73%, for *N. affinis* it is 18.51%, for *B. pervariabilis* it is 16.19% and for *Melocanna baccifera* it is 14.04%. Bamboo hemicellulose content usually refers to polysaccharides and carbohydrates. According to experimental observation, when polymer sugars, soluble sugars and starch contents reach to 20–30%, the bamboo will be vulnerable to insects or mildew attack. Through the catalytic function of hemicellulases, the hemicellulose in bamboo can be rapidly broken down to glucose, and in bamboo processing and utilization, insects and moulds are prominent issues, although there are now more matured technologies for the preservation of bamboo materials before processing.

LIGNIN CONTENT. Bamboo lignin content is generally 16–34%, and the content may vary in different species. For example, for Moso bamboo it is 26.41%, for *B. pervariabilis* it is 16.19% and for *Ph. glauca* it is 33.40%. The lignin content of bamboo also varies at different ages. Usually, with increasing age, the cellulose and hemicellulose content decreases year by year, while the lignin content increases. For Moso bamboo, after 6 years of age the lignin content stabilizes. This suggests that in the selection of bamboo materials for fibre making, younger bamboos are more suitable, e.g. for Moso bamboo, 3-year-old stands should be selected. In contrast, for the production of high strength products such as plywood, the Moso bamboo raw materials selected should be older, such as 5–6 years, because the lignin in bamboo tissues can enhance the strength of the cell walls and the bonding power of the fibres, which will strengthen the hardness of the bamboo material.

Fig. 2.50. Bamboo pulp oriented plantations (*Bambusa vulgaris*) in Brazil (Gutierrez-Céspedes and Mendes-Araújo, 2015).

Fig. 2.51. Bamboo fibre and products (Jin Wei).

Fig. 2.52. Bamboo fibre clothes (Jin Wei).

Fig. 2.53. Bamboo winding composite pressure pipes (Jin Wei).

2.4.2 Bamboo fibre utilization

Bamboo fibre utilization has a wide and surprisingly prosperous future. Towards the close of composing this book, the authors read about a report in the *China Economic Herald* (Zhou Haochen, 2017) that Zhejiang Xinzhou Bamboo-Based Composites Technology Co., Ltd. had developed a new material called bamboo winding composite material, which was the first ever in the world that uses bamboo as a material for various types of underground pressure pipes. According to the report, the material that the pipes were made of was a new bio-based composite that has certain comparative advantages in temperature tolerance, corrosion resistance and frost heave resistance, etc. over pipes made of traditional materials such as polypropylene (PP) fibre, glass fibre, metal or cement. The report also said that the material has advantages such as being lighter in weight, easier to install and with a longer shelf life. The material was introduced as an environmentally friendly product, which was very distinguished by its energy-saving and low-carbon production. For example, 10 million bamboo-based composite pipes could replace 33.6 million tons of steel pipes, which means replacing 45 million tons of crude steel, saving around 72 million tons of iron ore and reducing CO_2 emission by 82 million tons. The material could not only be used for underground pressure pipes, but also for large-scale underground utility tunnels. It was reported that the material was expected to be applied widely in rural housing, the cartridges of bullet trains, large-scale storage tanks, telegraph poles, etc. in the future. According to the report, the bamboo winding composite material also has comparative advantages in cost compared with traditional pressure pipes, and bamboo winding composite pipes may save 30% of the costs of these, and 60% of the costs of PP pipes. The author believes that this technology has opened a complete new frontier of bamboo application and bamboo industry.

2.5 Bamboos Play Important Roles in Supporting Human Lives and Have a Splendid and Long Cultural History

Bamboos have many characteristics ('spirits') that are particularly appreciated by humans. For example, they are evergreen, vigorous, and the shoots flourish in a short period of time after rain. They are tough and stand upright, yet flexible – they unify both principles and flexibility. Most of them are hollow in the centre, which represents their modality, but at the same time, their stems show their natural simplicity and elegance, while they have regular internodes; these characteristics represent austerity and integrity in personality. Bamboos are also like treasure trees, which dedicate everything to others without the requirement for a return.

These are the 'morals' or 'spirits' of bamboos that Chinese people appreciate. People in Asia and Latin America, and particularly in China, think that bamboos are far from purely plants of biological significance, but have the qualities of 'a person', or say that bamboos are personalized nature. Their characteristics represent the very basic and core parts – the feelings, ideas, thinking and ideals of Chinese people, and of other rich cultural heritages.

In China, the bamboo, chrysanthemum, plum and orchid are called the 'Four Gentlemen' (四君子) (moral models); bamboo, plum, pine and orchid are called 'the four friends' (best friends for people); bamboo, pine and plum are recognized as the 'Three Friends of Winter' (i.e. friends under severe conditions); bamboo, plum and stone (as in a piece of rock) are known as the 'Three Pure and Tranquil Friends' (and must be accompanied by a quiet and peaceful environment). So bamboos have often been praised by poets and writers as well as by the common people since ancient times. People expect to carry forward the bamboo spirit in human society. A large number of poets and scholars from ancient times have praised bamboo highly in their works, including poetry, paintings, signings and dances. In Asia, many parks, buildings, towns and villages, streets, tourist attractions, shops, restaurants, companies, etc. are named after bamboo, and many Chinese characters have the bamboo character as radicals. The famous British writer and scholar Joseph Needham (Noel Joseph Terence Montgomery Needham) even claimed that the 'East Asian civilization is none other than a bamboo civilization.' The influence that bamboo brings to human material life and spiritual life is called bamboo culture. Why does bamboo have such a special place in people's minds? It has to do with its contribution to humanity and the spiritual meaning that it embodies. Here we briefly introduce some information on spiritual aspects of bamboo that have existed in China since ancient times.

> **The British well-known scholar Joseph Needham says: 'The East Asian civilization is none other than a bamboo civilization'.**

2.5.1 China is a bamboo country with a long history

China was the first country in the world to utilize and cultivate bamboo

According to research, hominids 10,000 years ago living in the Yangtze River Basin had already started the cultivation and utilization of bamboos. They used bamboos for fishing, hunting and the production of tools and combat weapons. The Hemudu primitive society in Yuyao County, Zhejiang Province, which was as long as 7000 years ago, was found to have been using bamboos for baskets and other tools. Bamboo charcoal was found in the ruins of the Longshan Culture in Licheng, Shandong Province. More than 200 pieces of bamboo-made utensils were found in the Qianshanyang Relic Site of Wuxing, Zhejiang Province. These findings proved that bamboos were closely related to the everyday lives of the primitive society.

Bamboo was the first hunting and war weapon

The main weapon in ancient China was the bow and arrow. In 1977, the Chu (1042–223 BC) tomb in Hubei Province unearthed more than 450 kinds of weapons, most of which are bows with a bamboo arch or arrows with bamboo shafts. Since the invention of gunpowder, bamboo weapons made a further leap in technology. In AD 970 a rocket shaft made of bamboo was offered to the imperial palace of the Song Dynasty: with gunpowder primers tied to the bamboo shaft; by lighting the fuse, the burning gunpowder could be shot out through the bamboo tube, and provided a primitive weapon for war. During the Ming Dynasty (AD 1368–1644), bamboo strips were used to make 'Flying crows

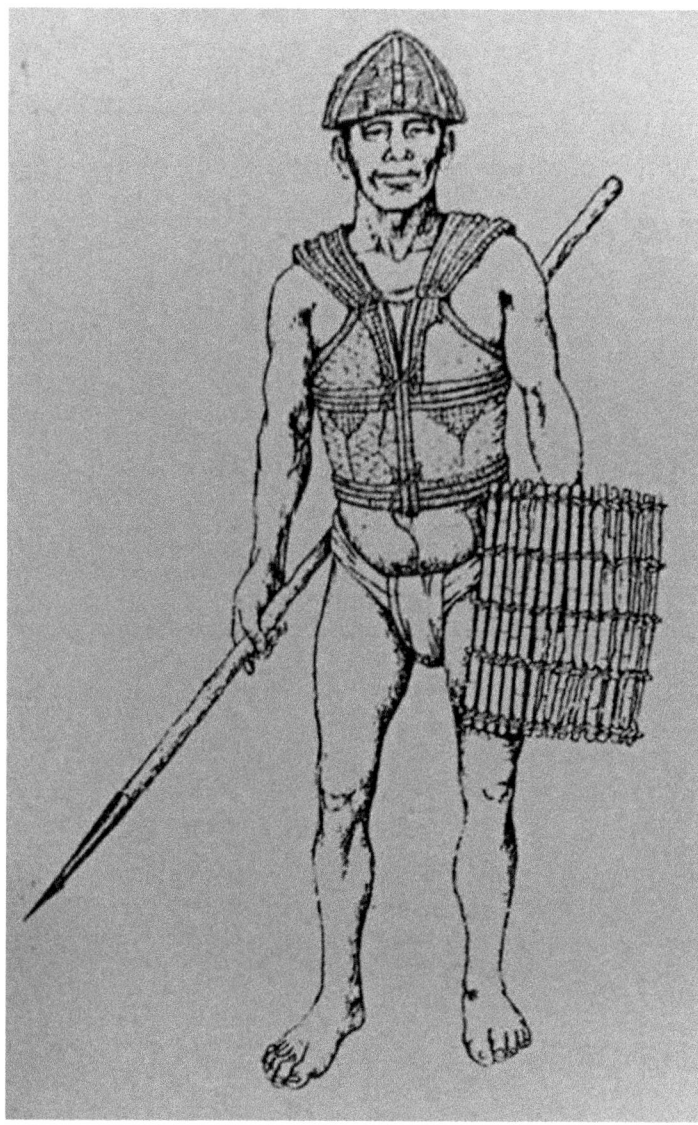

Fig. 2.54. An ancient Chinese man with bamboo products. Available at: http://sns.91ddcc.com/t/29682 (accessed 4 April 2017).

with magic fire' (神火飞鸦), gunpowder was filled in, and after being ignited, the crow could fly over 300 m, reach the camps and warships of the enemies and burn them. Two-stage bamboo 'rockets' appeared in the Ming Dynasty, and these used 1.5 m long bamboo cylinders.

Bamboo is closely related to people's lives

In the Western Zhou Dynasty (1046–771 BC), bamboo walls were an important part of residential housing. In the Warring States period (770–225 BC), bamboo articles became an important industrial and productive sector of the society. By then, bamboos had already become indispensable survival elements for human beings. The Han Dynasty (202 BC–AD 220) had more than 60 types of different bamboo utensils and tools, the Jin Dynasty (AD 265–420) had more than 100, the Tang Dynasty (AD 618–907) had nearly 200 and by the time of the Qing Dynasty (AD 1636–1912),

Fig. 2.55. Bamboo plaited pastry box in the Ming Dynasty (left) and bamboo plaited bookcase in the earlier Song Dynasty (Zhu Zhaohua).

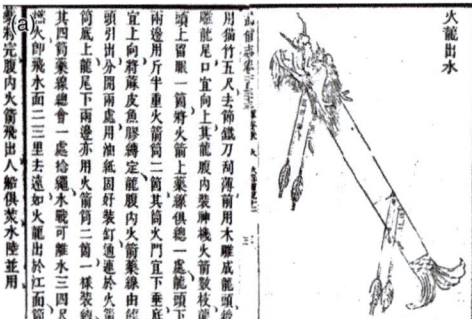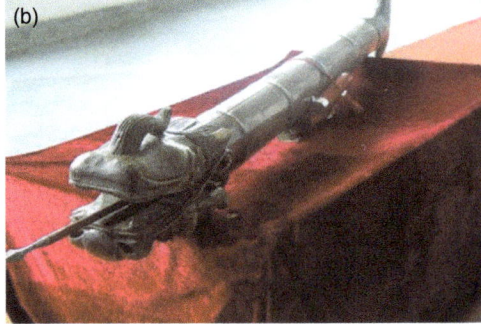

Fig. 2.56a,b. Bamboo weapons from ancient China. Available at: http://derek.chen.blog.163.com/blog/stat ic/11893148200722684652253/ (accessed 4 April 2017).

there were more than 250. These bamboo-made products were included in all aspects of people's daily life:

> For example, in cooking utensils, there were baskets, bowls, spoons, plates, steamer[s], bookcase[s], trunks, boxes, piggy banks; in furniture, there were beds, couches, seats, chairs, pillows, screens, tables, cabinets, counters; in counting tools, there were counting rods, abacus[es]; for measurements, there are rulers, tubes; for lighting, lanterns, candle cabinets; for sanitation, brooms, smoking cages; for decoration, curtains, flower vases; for appreciation, toys, fans, walking sticks; for gambling, gambling chips; for funerals, bamboo coffins. The applications of

bamboo are a major feature of ancient Chinese civilization, and is [are] closely and historically related with the heritage and strengthening of the Chinese nationality, it [they] could be regarded as a major sign of the Chinese civilization.

> (Wu Danren, 1999, p. 5.)

Bamboo played an important role in agricultural development in China

Bamboos were important materials for agricultural production tools, irrigation, drainage and fishing. Agricultural production tools made of bamboo include flails, rakes, grain winnowers, screeners, shallow baskets and

Fig. 2.57. Bamboo baskets for rice harvesting (Museum of Bamboo Weaving Articles in Qingshen, Sichuan Province, China).

mats for grain drying; grain storage facilities, such as huge woven baskets that have square bottoms and round tops; farm tools and irrigation tools, such as bamboo taps, bamboo tunnels and cylinder cars; and fishing tools such as bamboo traps, fishing rods and cages made of bamboos. In the Spring and Autumn Period and the Warring States Period in China's history, our ancestors created a leverage tool for drawing water from underground called a 'JU', as well as an irrigation facility, 'the high-speed winder'.

China's great ancient water conservancy project, Dujiangyan, in Sichuan Province, was designed by the famous officer Li Bing and his son (250 BC). They had an enormous amount of large bamboo cages made and

pebbles were put inside them; the cages containing pebbles were piled up and formed a levee in the middle of the river, which split it into two and channels and successfully achieved the current distribution of the large river – the Min River (or Minjiang, a tributary of the Yangtze River). The bamboo cages with pebbles inside played an important role in this and were one of the key successful technologies of this ancient hydro-project. This levee, after 2260 years, is still working well and playing an important role in drought and flood control and irrigation for croplands. The Dujiangyan hydro-project brought irrigation to 36 counties (cities, districts) in seven prefectures, with a total irrigated area of 669,000 ha, which has now become one of the famous

Fig. 2.58a,b. High-speed winders (irrigation water wheels) made of bamboo (Museum of Bamboo Weaving Articles in Qingshen, Sichuan Province, China).

fertile granaries of China. Dujiangyan is the world's oldest and only well-conserved large-scale ancient hydro-project that realized non-dam water diversion, and has been chosen as a World Cultural Heritage site.

Fantastic uses of bamboo
in the mining industry

Let us take ancient coal, salt and gas exploration as examples. China began large-scale coal mining

Fig. 2.59. Bamboo cages form a traditional Chinese levee. Available at: http://dingqiya.com/?p=74 (accessed 4 April 2017).

2500 years ago, and as early as the Western Han Dynasty (202 BC–AD 9), in Song Yingxing's ancient book *Wonderful Creations* (*Tian Gong Kai Wu*, 《天工开物》), there was a record of using bamboo tubes for removing the harmful gas in mining caves.

There are also records of large-scale salt mining in China 2000 years ago. This was documented in the 3rd century BC in a book called the *Chronicles of Huayang* (*Hua Yang Guo Zhi*, 《華陽國志》), which records the case of the salt exploitation in Sichuan. Even at the very beginning of salt-mining projects, bamboos were already playing key roles, and a series of ingenious bamboo tools were designed to draw salt from salt wells hundreds of metres deep. Basically, bamboos were involved in three of the main processes of salt mining:

1. There was a bamboo fan piston cylinder for sinking pits into the ground. Iron 'vertebrae' wrapped in bamboo cables were used for breaking the hard parts underground. When the bottom of the pit reached 7–10 m below ground level, there were many types of gravel that hindered further sinking of the pit and the bamboo fan piston cylinder was then used to take out the gravel.
2. When the salt well had been made, the next step was drawing brine, and ancient people made an ingeniously designed bamboo brine drawing canister, which used cooked leather as its piston.

3. Before the advent of metal and chemical pipes, bamboo culms were used to transport the drawn brine. This transport facility made of bamboo was called a 'Zhu Jian'. Large-sized bamboo culms were used to build the facility. The inside nodes of the culms were removed, so that they became large tubes, and these large tubes were then connected with each other; sometimes, the total length of the bamboo transmission pipeline could reach thousands of or tens of thousands of metres. In the production area, the pipes were arranged in criss-cross patterns, which made a very spectacular scene.
4. The 'Zhu Jian' can also function as a transmission pipeline for natural gas.

The 'Zhu Jian' was used in China 1800 years ago for the mining of natural gas, and was still made of bamboo. In the Ming Dynasty, in the famous *Wonderful Creations* (fifth volume), it was recorded in detail that bamboo was used to transport natural gas. In AD 1782 (the Qing Dynasty), large amounts of natural gas were mined from deep rock and high pressure areas underground, and dozens of 'Zhu Jians' were usually used at the same time for transmission of the gas. Innumerable bamboo tubes were connected and made into pipes. The connection technology was exquisitely structured. In a book called the *Artesian Well* (*Zi Liu Jing Ji*, 《自流井记》), written by Li Rong in the period of the Guang Xu Emperor in the Qing Dynasty, there was a detailed description of the technology used: 'The pipes are wrapped with bamboo strips, hemp was used to wind outside of the bamboo strips, apply putties and make them infiltrate; therefore, the connection is very tight, and no raining water could filtrate, nor any leakage emerge from inside. Each pipe could reach thousands of steps away.'

China is the first country to carry out a comprehensive study on bamboo

The Book of Songs (*Shi Jing*, 《詩經》), which was estimated to have originated around the 6th century BC, *The Classic of Mountains and Seas* (*Shan Hai Jin*, 《山海經》) in the 4th century BC and *The Tribute of Yu* (*Yu Gong*, 《禹貢》), which is China's oldest known book of geography in the 3rd century BC, have all recorded and described China's bamboo species, distribution, uses and economic values. During the Western Han Dynasty, Sima Qian in *Historical Records* (*Shi Ji*, 《史記》) (104–91 BC), mentioned for the first time a special government official position set up to manage bamboo. In Gu Sixie's *Arts for the People* (*Qi Min Yao Shu*, 《齊民要術》)

from the 5th century there was a systematic introduction to the technologies of bamboo afforestation, bamboo shoot processing and preservation. After the Jin Dynasty, many monographs about bamboo came out; these include *Zhu Pu* (*The Bamboo Monograph* 《竹谱》) by Dai Kaizhi (AD 520–579), and the *Bamboo Shoot Monograph* (*Sun Pu*, 《筍譜》) by Zan Ning (赞宁, AD 919–1001) of the Song Dynasty. The latter book gave a detailed introduction to the processing of and preservation technologies for bamboo shoots, as well as the cooking methods. Wang Zhen of the Yuan Dynasty (about AD 1271–1368) wrote the *Wang Zhen Agricultural Book* (*Wang Zhen Nong Shu*, 《王禎农书》), which has very detailed records of bamboo cultivation, flowering and fruiting, harvesting, the digging of bamboo shoots, bamboo forest fertilization, etc. This book has made an outstanding contribution to the study of bamboo. The *Compendium of Materia Medica* (*Bencao Gangmu*, 《本草纲目》) of 1590, a book by Li Shizhen of the Ming Dynasty, has detailed descriptions of the bamboo species that can

Fig. 2.60. Salt mining using bamboo tools (from the Museum of Salt Exploration in Zigong, Sichuan Province, China).

Fig. 2.61. Gas transfer using bamboo pipes (from the Museum of Salt Exploration in Zigong, Sichuan Province, China).

be used for medicines, including their growth environment, morphology, medicinal parts (leaves, roots, mushrooms (mycorrhizas), sheaths and stems) and their characteristics and tastes, and the major functions, dosage, usage, etc. of these medicines. In *Florilegium* (*Qun Fang Pu*, 《群芳譜》) by Wang Xiangjin of the Ming Dynasty, there was detailed introduction to the technologies of bamboo cultivation, silviculture and transplanting, as well as bamboo forest management.

These technologies are still the principles of today's bamboo cultivation and management. Why did the ancient Chinese pay such a lot of attention to the study of bamboos? Because they have made tremendous contributions to human beings, and have huge value for and profound impacts on not only the material aspects of life, but also the spiritual aspects.

Bamboo contributes to both the rich and the poor and provides convenient construction materials for resident housing, leisure and offices

Bamboo will become an important new construction material that is welcomed by people all over the world in the near future.

Bamboo constructions have been a unique architectural form since ancient times (Villegas, 2003). They can be very simple and cheap structures that provide accommodation for the poor. They can also be magnificent palace buildings in various forms and styles, such as pavilions, temples, and so on. Emperor Wu of the Han Dynasty (156 BC) constructed a bamboo palace near the Sweet Spring Temple (an ancient temple located

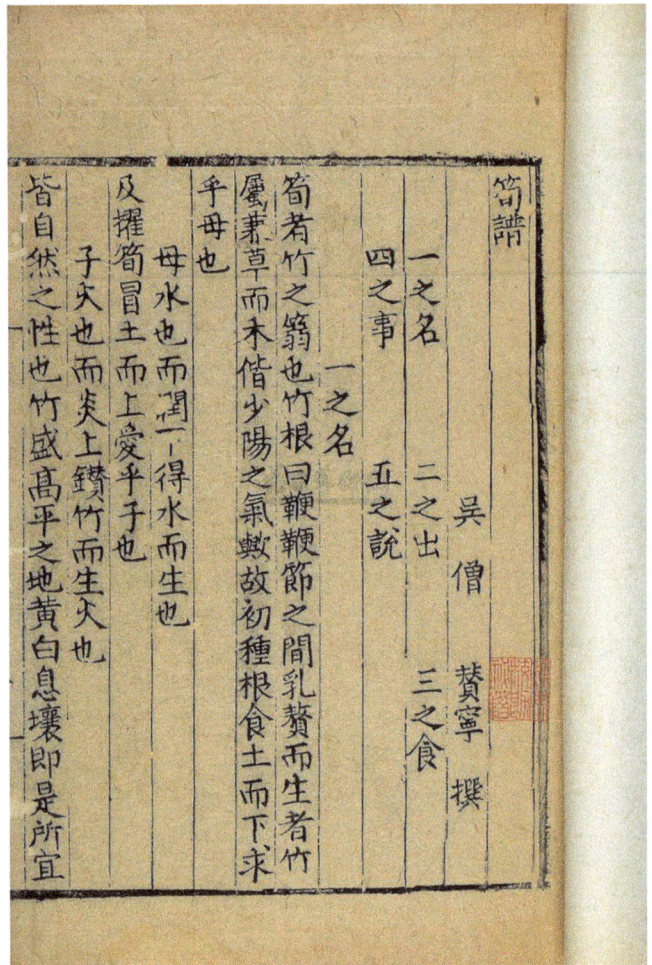

Fig. 2.62. First page of book the *Bamboo Shoot Monograph (Sun Pu,《筍譜》)* by Zan Ning (赞宁, AD 919–1001) of the Song Dynasty. Available at: http://auction.artron.net/paimai-art39962029/ (accessed 4 April 2017).

in Zaozhuang, Shandong Province), which was so named because of the sweet spring that was found inside the Temple. There is also an ancient Gingko tree that is estimated to be more than a 1000 years old in the temple, which provides evidence of its long history. Literature of the age recorded that: 'There is a palace made of bamboo, where the Son of Heaven lives in. The Bamboo Palace is great and beautiful, standing on the right side of the Sweet Spring, while the bamboo hall is high and wide, with Gods and Fairies by its side (以竹为宫，天子居中。以竹宫丰丽于甘泉之右，竹殿弘敞与神嘉之傍。)' so it is obvious that the bamboo palace was very luxurious by then. Currently in China, parks, tourist attractions and luxury hotels are almost all decorated with bamboo structures or have bamboo structural materials that provide and promote the cultural tastes of the buildings and environment.

Bamboo buildings have long histories in Asia, Latin America and Africa. When travelling in East Africa and West Africa, as well as in Latin American countries such as Bolivia and Ecuador, I saw a lot of poor rural people living in very simple houses with frames made of bamboo poles and woven mats, while the outside was pasted with mud mixed with grass. In Ecuador, a charity factory established by the church Hogar de Cristo has been using bamboo to build temporary houses for the victims of disasters or homeless people. In some Asian countries, such as the Philippines, Vietnam and Laos, and in many Latin American countries, there were many rich people who liked to live in bamboo houses. In these countries, there were a lot of bamboo hotels, restaurants, schools (such as the University of Construction Engineering in Colombia), churches (in Ecuador, Colombia, Bolivia and the Philippines), office buildings

Fig. 2.63. A bamboo pavilion in Hangzhou, Zhejiang Province, China (Zhu Zhaohua).

Fig. 2.64. The office building of Colombia's Ministry of Environment, which is mainly made of bamboo (Zhu Zhaohua).

Fig. 2.65. The internal appearance of the office building of Colombia's Ministry of Environment, which is mainly made of bamboo (Zhu Zhaohua, 2005).

Fig. 2.66. A bamboo house in Colombia (Zhu Zhaohua, 2005).

Fig. 2.67. A bamboo cottage and bamboo baskets in Ghana (Zhu Zhaohua).

(for example the office building of Colombia's Ministry of Environment), shops, conference halls (in Colombia and Bolivia), apartments, petrol stations, highway toll stations, stables and milk plants, and so on, all made of bamboo construction. I saw someone trialling the use of bamboo for the frames of bicycles and jeeps in Colombia (Villegas, 2003; see Figs 2.73 and 2.74).

Since the 1990s, due to the invention and application of laminated bamboo materials, bamboo curtain and mat boards and pressed bamboo materials, as well as the mass production of the laminated and pressed bamboo materials in China, laminated and pressed bamboo materials have started to be used for construction on a large scale. For example, in 2010, China's bamboo curtain/mat board production reached 3.58 million tons and the production of laminated bamboo flooring reached 1.11 million m³. Now more than 85% of China's cement moulding board is made of bamboo. Since the 'Reform and Opening-up Policy' in 1978, bamboo has contributed significantly to unprecedentedly large-scale construction projects in China. At the same time, bamboo laminated materials have been largely applied in construction as structural parts, such as beams and columns, as well as being used as interior decoration and furniture panels, veneer materials, etc. Also, pressed bamboo technology has turned bamboo slices into lumber materials, which can widely replace (tree) timber materials in furniture manufacture and construction. For example, in the Wuxi Conference Hall – China's third largest conference hall – which was launched for use in 2008, almost all of the interior decorative materials used pressed bamboo, including the walls, ceilings, floorings and furniture. From the above-mentioned new developments, we may predict that the future will see more and more people welcoming various different types of natural, environmentally friendly and beautiful bamboo construction materials, structures and buildings into their lives.

Transportation facilities made of bamboo are very popular

Transportation facilities made of bamboo were very popular in ancient China and some of them

Fig. 2.68a,b. Inside appearance of a church in Leyte in the Philippines, which is mainly built with bamboo (Zhu Zhaohua).

Fig. 2.69. Bamboo furniture made from *Phyllostachys arcana* (INBAR).

are still being used today. For example, the single-rope bridge, double-rope-bridge, multi-rope suspension bridge, multi-rope walking bridge and various types of bamboo structural bridges. There were also ancient water transport structures made of bamboo, such as bamboo rafts and boats, and transport facilities in the mountains such as bamboo sedans. They are all now still in use, and are commonly seen in tourism sites. Emperor Ming (AD 28–75) in the Han Dynasty built a famous bamboo rope bridge over the ferocious river of Lancang. The famous

bamboo rope suspension bridge over the Min River is 230 m long and 3 m wide; each rope is 20 m long and their circumference is over 50 cm; there are handrails on both sides; the whole bridge is made of bamboo. Bamboo bridges are also popular in Colombian and Chinese gardening.

Bamboo rafts, used as ferries to cross bodies of water, have a long history. In ancient times, these rafts were used to transport soldiers in wartime. As described in the *Heroical Biography* (《汉末英雄记》), Cao Cao (曹操) (155 BC–AD 220) sent his troops to the side of the Yangtze

Fig. 2.70. School desk made from bamboo (INBAR).

Fig. 2.71. A church built with bamboo in Colombia (Simón Veléz, Colombia).

Fig. 2.72. A bike made of bamboo in Colombia (Zhu Zhaohua).

Fig. 2.73. A jeep made of bamboo in Colombia (Zhu Zhaohua).

River, and they were intended to cross the river at Chibi, but there were no boats available, so he then decided to use bamboo rafts as transport for the troops.

Records of bamboo boats have been found from as early as in the Warring States Period (475–221 BC). Bamboo warships were also built in the Southern Dynasties (AD 420–589).

Fig. 2.74. A bike made of bamboo (*Bambusa vulgaris*) in Ghana (Zhu Zhaohua).

Fig. 2.75. A bike made of bamboo (*Fargesia* sp.) in Yunnan Province, China (Zhu Zhaohua).

In 2012 and 2013, when I was visiting the Philippines and Timor Leste, I saw some very beautiful bamboo ferries and fishing boats. In the middle 1990s, Hunan University of Science and Technology in China produced a very light and strong mini-cruise boat with bamboo mats and plastic steel. From the 1990s, China started to use bamboo and wood composite laminated panels for the lining panels and floors of train and truck containers. To date, China is the biggest producer of containers in the world, and the use of bamboo in container production has opened up a new field of application for this wonderful natural material.

The bamboo sedan is a unique man-powered transport facility in China. In the Warring States period (770–221 BC) it was already very popular, especially in the southern mountainous areas, and it became a commonly used means of transport for officials and noblemen. The famous Tang Dynasty poet Tao Yuanming (陶渊明) (AD 365–427) mentioned the comfort of bamboo sedan: 'I had always been tortured by my foot disease, I always choose to take the bamboo sedans, because it felt very comfortable (素有脚疾,向乘篮与,亦是自适).' Now, in tourist spots in mountainous areas of China, such as E'mei Mountain and Qingcheng

Fig. 2.76. A bike made of bamboo veneer in Zhejiang Province, China (Zhu Zhaohua).

Fig. 2.77. A bamboo raft and bridge in Zhejiang Province, China (Zhu Zhaohua).

Fig. 2.78. A bamboo raft in Sichuan Province, China (Photo provided by Chen Yunhau).

Fig. 2.79. A boat made of bamboo panels in South Korea (Zhu Zhaohua).

Mountain in Sichuan Province, Yellow Mountain in Anhui Province and Zhangjiajie in Hunan Province, visitors can still enjoy the use of comfortable bamboo sedans.

Food and medicine

The relationship between bamboo food and people can be described as 'No bamboo shoot,

Fig. 2.80. Boats made of bamboo and with bamboo woven sails (from the Museum of Bamboo Weaving Art, Qingshen, Sichuan Province, China).

Fig. 2.81. A bamboo sightseeing ship in the Philippines made with 70% of materials from bamboo (Zhu Zhaohua).

no dinner (无笋不成席).' Shoots are an obvious portion of the Chinese diet. There are a great number of poems and articles praising shoots in the history of China. During the Tang Dynasty, Bai Juyi, a member of the literati, suggested that bamboo shoots were as important as fish. He said: 'Fish and bamboo shoot made me full in the morning, the summer cloths made of banana yarn is so light (鱼笋朝餐饱,蕉纱暑服轻。).' The famous writer of the Song Dynasty, Su Che, expressed his indifference to fame and wealth by writing in the poems that 'After collecting the tea, steam the yellow leaves outdoors, dig shoots in the forest and cultivate green vegetables. I've fully felt the true saturation of the life, and am trying to build a humble house

Fig. 2.82. The floor of a container built with bamboo (Zhou Guomo and Shi Yongjun, 2012).

beside the monks' temple (来摘茶户外蒸黄叶，掘笋林中间绿蔬。一饱人生真易足，试营茅屋傍僧局。).' These two poems expressed that the poets enjoyed a life with bamboo shoots.

The ancient Chinese have done in-depth studies on different bamboo species that could be used for food purposes. For example, the Song Dynasty monk Zan Ning described 98 bamboo shoot species that included 32 species used for edible shoots (Fig. 2.62.). In China's long-established history of bamboo utilization, people formed a complete set of gastronomic techniques with bamboo shoots, which could produce dynamic types of dishes and various flavours in different locations. In 1997, during the first China Bamboo Cultural Festival, the host County – Anji County of Zhejiang Province – prepared a 'Banquet of a hundred bamboo shoot dishes' for the appreciation of the guests and friends of the festival. Fresh bamboo shoots can be cooked directly as a fresh vegetable, or they can be processed into dried/baked shoots, shredded shoots, fermented shoots, and boiled and canned shoots. The canned shoots can be stored for a comparatively long time (more than a year). In addition to the above, Chinese ethnic minorities have their own unique bamboo shoot products, such as 'stinky shoots' and 'sour shoots', which are fermented shoots, as well as bamboo rice, which are all very special folk foods. Statistics show that in 2010, China's production of fresh shoots was 6–7 million tons, among which 1.6 million tons were processed. In addition to bamboo shoots, bamboo rice is also one of the other nutrient-rich foods that bamboo has contributed to humanity.

The medicinal values of bamboo have long been recognized by Chinese people. Bamboo juice (*Succus Bambusae*), the green part (bamboo bark) and the yellow part (*Tabasheer*) of the bamboo culm, and bamboo shavings or 'Zhu Ru' (*Caulis Bambusae in Taeniam*) are all included in the *Chinese Pharmacopoeia*.

Bamboo and cultural goods

It is universally known that the brush pen is a traditional Chinese writing tool, and that the pole of the brush and traditional penholders are both usually made of bamboo. Before the invention of paper, bamboo slips were an important tool for inscribing Chinese characters. The earliest Chinese characters were seen on the oracle-bone inscriptions in the Shang Dynasty

Fig. 2.83a,b. Bamboo shoot dishes (Anji Forestry Bureau, Zhejiang Province, China).

Fig. 2.84a,b. A bamboo culm banquet in South Korea (right). Food is placed in a piece of culm of Moso bamboo and put into a steam heater (left) (Zhu Zhaohua).

(1600–1046 BC), where the characters were carved on tortoise shells. Because the oracle bones were of different sizes, it was difficult to bind and store them. In the 3rd century book *Records of Zhou* (*Zhou Ji*, (《蔡伦》) 《周记》), it was recorded that bamboo slips were made specifically for writing texts, and that the length of each slip was 40 cm; 40 characters could be written on each piece of bamboo with a brush. Written or inscribed bamboo slips that were strung together by leather cords or ropes were the first books in China. It was recorded that Qinshihuang, the first emperor of the Qin Dynasty, read 150 kg of bamboo slips with memorials written on them every day. China's most famous military strategy book, Sun Tzu's *The Art of War* (《孙子兵法》) was excavated from one of the tombs found at Yinqueshan in 1972, inscribed on bamboo slips. The tombs date from the Han Dynasty (206 BC–AD 220) and the bamboo slips were found to include two separate texts – one attributed to Sun Tzu – *The Art of War* and the other to his descendant Sun Bin (*Art of War*, 《孙膑兵法》), although these may be overlapping texts. The present Yinqueshan Han

Tomb Museum in Shandong Province still holds these two pieces of work.

In addition to the Chinese brush and penholders mentioned above, bamboo stationery includes pen caps, pen racks, bamboo ink stone holders, rulers, bamboo measuring tapes, balance boxes and so on. Further, in the East Han Dynasty, Cai Lun (蔡伦) (AD 50–121) invented papermaking technology. Because the fibre content of bamboo is up to 40%, it became one of the important raw materials for papermaking. In the Jin Dynasty, bamboos started to be used by Chinese people for papermaking – 1000 years earlier than Europe. The utilization of bamboo has greatly promoted the development of the paper industry, and the prosperity of Chinese culture. In India, the bamboo paper industry enjoys a long history and has also reached a considerably large scale. Now Brazil and Vietnam also have quite large scales of bamboo paper industries.

From the above brief introduction, it may not be difficult to understand the enormous contributions that bamboos have made to the material civilization of human beings, and why

Fig. 2.85. The bamboo slips recording the book *The Art of War* excavated in 1972 from a tomb found at Yinqueshan dating back to the Han Dynasty (206 BC–AD 220), and presently kept in the Yinqueshan Han Tomb Museum in Shandong Province (see text for further details). Available at: http://pyzhaomin.blog.163.com/blog/static/15524297520111074754938/ (accessed 4 April 2017).

bamboo is the favourite of so many ancient literati and a broad mass of people. Here, allow me to summarize using a paragraph written by Su Dongpo, a great member of the literati and poet of the Song Dynasty, when he was praising bamboos for their contributions to the people in his home town – Meizhou (now Meishan) in Sichuan: 'people eat bamboo shoots, live under bamboo tiles, transport with bamboo rafts, use bamboo for fuels, wear bamboo clothes, write on bamboo papers and walk with bamboo shoes (食者竹笋,庇者竹瓦,载者竹筏,炊者竹薪,衣者竹衣,书者竹纸,履者竹鞋。)'. In this extract from Su Dongpo, we can see that bamboo was related to all of the necessities of people's lives in his home town. No wonder this famous poet created so many well-known poems and paintings about bamboo, he had extremely deep feelings for bamboo.

Fig. 2.86. A modern bamboo slip book of *The Art of War* (see text and Fig. 2.85 for further details). Available at: http://www.360doc.com/content/12/0804/14/4310958_228298342.shtml (accessed 4 April 2017).

Fig. 2.87. Bamboo culms being transported on the river in Anji, Zhejiang Province, China (INBAR).

2.5.2 Bamboo and spiritual civilization

Asia, Latin America and Africa all have colourful bamboo cultures. Bamboos represent the human spirits that are commonly appreciated by many cultures and have made a great contribution to the material civilization of human beings. Bamboos, in people's minds, have fine connotations that are reflected in the aesthetic interests, religious spirits, values and personality ideals of people.

> *Bamboo embodies and symbolizes the fine human spirits.*

In the pursuit of good and beautiful things in real life and spiritual life, people often tend to use things that they admire, including plants, animals and all other things existing in nature, to make a comparison. For example, when describing somebody who is not afraid of hard conditions, Chinese people usually use the Chinese plum (*Prunus mume*) as a comparison, because it blossoms in the cold wintertime and is a harbinger of spring, and the flower colour is usually red (Plums blossom in the Spring, the flowers are usually red. Winter sweet (*Chimonanthus praecox*), which also blossoms in the winter, and usually has yellow flowers, is seen as a symbol of endurance in severe winters. There are also comparisons for bad spirits. For example, when describing a person who is cunning, people usually describe him/her as a fox; for somebody who is lazy, people compare him/her with a pig.

Bamboo, in the eyes of Chinese people, is the symbol of the ideal Chinese personality, with almost all fine human spirits and habits in one, as described in the list below:

1. Evergreen, beautiful posture and a peaceful, harmonious atmosphere: bamboo forests and bamboo gardens have always been the residence

Fig. 2.88. A village in an area of bamboo in Anji, Zhejiang Province, China (Zhu Zhaohua).

places of the Chinese literati, scholars and artists in history. Many Buddhist shrines were built beside or in bamboo forests, and temples were commonly seen in bamboo forests in many places in China. The famous poet of Yangzhou, Luo Pin (AD 1733–1799) describes the comforts of living in a bamboo forest as follows: 'In the bamboo forest the breeze is clean and refreshed, no matter how dusty it is outside of the forest, though the wind continues to blow, there is little dust here. There are not many places that are as clean as here, I just hope to ride with an eminent monk on the crane (竹里清风竹外尘，风吹小到少尘生。此间干净无多地，只许高僧领鹤行。).' In the Jin Dynasty, there were the 'Seven Sages of the Bamboo Grove' (竹林七贤), who gathered in the bamboo grove often to enjoy a retreat or celebrate nature. In the Tang Dynasty, there were the 'Six Escapists of the Bamboo Stream' or 'zhúxīliùyì' (竹溪六逸) represented by the famous poet Li Bai (AD 701–762), who secluded themselves in bamboo forests to drink and write poetry. Where bamboos are present, people always feel the refined, noble spiritual interests. Several stands of bamboo will be able to give a vibrant embellishment to a small farmhouse, or add elegance to a luxury hotel. Bamboos provide people with leisure, comfort and a harmonious environment, and they have become important elements in the development of China's ecological tourism. Some famous tourist attractions in China feature bamboo, for example, the Shu Nan Zhu Hai (the Bamboo Sea in South Sichuan) in Sichuan Province. The top 'China Bamboo Hometown' – Anji County in Zhejiang Province – had 10.44 million tourists in 2013, and the total ecotourism income reached 4.09 billion CNY; in the same year, the 800 farm-stay hotels managed by bamboo farmers achieved a total income of 60 million CNY. The majority of Anji's tourist attractions feature bamboo.

2. Modesty and integrity: bamboo is hollow inside and has sequential internodes. These features match with the spiritual principles that Chinese people regard as noble, such as being open-minded to others, being modest and learning from others, and keeping to principles (having integrity). These principles are considered to be the personalities of the noble. Bamboo contributes to people of all levels of the society without distinction of their lowliness and nobleness. They do not grumble when they are made into brooms and held in the hand as a cleaner, nor feel pride when they are brush pens in the hands of a painter. Bamboos are willing to sacrifice to all who love them. Bamboo's modesty, spirit to move onward and upward, and perseverance are often used to describe noble persons with perfect characters – 'All Virtuous Gentlemen'. The Tang Dynasty scholar, Bai Juyi, said: 'The water is light, it may help me to lift my mind higher and out of the dust of chasing fame and wealth, water is my friend; bamboo is hollow inside, which helps me to keep modest, bamboo could be my teacher (水能性淡为我友，竹解心虚是吾师。).' The Chinese Painting Master Li Kuchan (李苦禅) once made a tribute to bamboo saying that 'The nodes (principles) are already there before bamboo is out of the ground, however when it reaches to the sky, it is still hollow inside (modest) (未出土时便有节，及凌云处尚虚心。).'

3. Fortitude, uprightness and vitality: bamboo can adapt to very harsh environments, even the winter frost, and remain lush, so bamboo, together with pine and plum are called the 'Three Friends of Winter'. Bamboos have strong adaptability and vitality, they grow fast. When Chinese people describe a person or a thing that develops with great exuberance and speed, they always use the phrase 'bamboo shoot after rain'. Bamboo is tough, unbending yet flexible. These are the perfect traits of personality that people appreciate. Bai Juyi, in his *Notes of Bamboo Raising* (《养竹记》), used bamboo to describe the ways of establishing one's virtuous principles:

> The nature of bamboo is fortitude, it is the base for establishing one's virtuous principles, a noble person who sees this character, will think about holding to the virtues and do not change; the nature of bamboo is uprightness, it is the foundation of a noble's stand, seeing this nature, a noble will think about keeping stands, not to be deviated; the nature of bamboo is hollow inside, when one's mind is 'hollow' (clear), he or she is able to understand the 'Way' (nature's course), when a noble sees this nature, he or she will be open-minded when seeing new things; bamboo's nature is to have nodes, the nodes represent integrity, when a noble sees the nodes, he or she will make efforts to encourage faithfulness to principles, always keep to the principles no matter in danger or peace (竹本固，固以树德，君子见其本，则思善建不拔者，竹性直，直以立身。君子见其性，则思中立不倚者。竹心空，空以体道，君子见其新，则思应用虚受者。竹节贞，贞以立志。君子见其节，则思砥砺名行夷险一致者...).

4. Harmony, coexistence and common prosperity: a bamboo grove is composed of major parts above ground (culms, branches and leaves) and major parts underground (rhizomes, shoots and roots). Rhizomes of different ages nourish the next generation together (the bamboo buds develop into shoots). One rhizome connects different generations (ages) of bamboo stands; they grow harmoniously together, sharing nutrients with and relying on/supporting each other. Despite arbitrary harvesting by human beings whenever needed, bamboos keep their sheer tenacity in flourishing from generation to generation. They are one of the large-sized plants that have the strongest self-renewal capacity and one of the easiest forest plants to manage. They are of high value in both economic and human spiritual life. Li Fang (AD 925–996) said bamboo has the four virtues of 'fortitude, gentleness, loyalty and righteousness

(刚、柔、忠、义)'. Among the many historical poems and articles in praise of bamboo, the most widespread is the poem by the Song Dynasty poet Su Dongpo (AD 1037–1101): 'With Yuqian Monk in Green Veranda (《於潜僧绿筠轩》).' In the poem, Su Dongpo places the material and spiritual connotation of bamboo through a profound revelation:

> I would rather to live without meat but with bamboo. Without meat, one may get thin, without bamboo, one will get philistine. A thin man could get fat again, but a philistine man may not be cured. The others may laugh at what I say, is this wise or stupid. How can one enjoy bamboo at the same time being bourgeois, is there such a thing in the world that can make 'incompatible goals' work? (可使食无肉,不可居无竹。无肉使人瘦,无竹令人俗。人瘦尚可肥,士俗不可医。旁人笑此言,似高还似痴。若君对此君仍大嚼,世间那有扬州鹤。).

I would rather to live without meat but with bamboo.

Fig. 2.89. The Chinese poet Su Dongpo (AD 1037–1101). Available at: http://auction.artron.net/paimai-art5025430036/ (accessed 4 April 2017).

2.5.3 Bamboo embodying and influencing Chinese culture

Bamboo elements have always been deeply rooted during the development of Chinese culture and has had profound impacts on almost all aspects of the culture (in characters, literature, painting, music, dance and craft arts). Therefore, a unique form of culture for the Chinese bamboo was developed.

Chinese characters

Ever since ancient times, there have been different types of daily used utensils, tools or artwork that are made of bamboo. As a result, the radical (root) character representing bamboo, 竹, is very popular as a component of many Chinese characters. Tables 2.17 and 2.18 demonstrate that bamboo plays (and has played) significant roles in agriculture, industry, culture and people's daily lives over the thousands of years of Chinese history.

Literature

In addition to the world's first bamboo monograph, *The Bamboo Monograph*, China has quite a lot of published literature, including a number of records and books, on the cultivation, taxonomy and applications of bamboo. In almost all of the dynasties of China's history, there emerged enormous numbers of bamboo-related poems and literary works. Most of these are anecdotes on bamboo, in which the literati praised the graceful outlook and superior virtues of bamboos.

The oldest Chinese poetry anthology, the *Book of Songs*, collected five poems about bamboo. Many Tang Dynasty poets enjoyed bamboo and chanted for it. The most representative of these included Du Fu, Wang Jian, Yao Han, Bai Juyi and Wang Weizhu (杜甫，王建、姚含、白居易、王维诸). Emperor Tai Zong of the Tang Dynasty also praised bamboo in his poems. Su Dongpo, a poet of the Song Dynasty, and Zheng Banqiao (郑燮) (1693–1765 AD), a prominent artist from the Qing Dynasty were both well known for their bamboo-related works in Chinese history. Zheng Banqiao was good at bamboo paintings, and produced over 100 inscriptions on paintings, which were quite diversified and incisive. In his painting *Bamboo and Rocks* (竹石) (see Fig. 2.90b), his inscription on top was a poem: 'Hold tight to the green mountains and never let loose, the bamboo establish their stands in broken rocks. Assailed by tens of thousand times of hits, they still stand there, no matter how the wind blows (咬定青山不放松,立根原在破岩中。千磨万击还坚劲,任尔东西南北风。).' The meaning here is "Went through much suffering, experienced so many hardships, the bamboo still hold its position and be brave and strong, without any fear."

Table 2.17. Statistics on the development of bamboo radicals (Wu Danren, 1999).

Year	Source (reference)	Total number of characters	Number of characters with bamboo radicals
Yin and Shang Dynasties (16–11th century BC)	Oracle bone inscriptions (甲骨文)	2700	6
Zhou Dynasty (11th–3rd century BC)	Inscriptions on bronze objects (金文)	2700	18
East Han Dynasty (AD 121)	*Shuo Wen JieZi* (说文解字), the *First Dictionary of Chinese Characters*	9353	151
Liang Dynasty (AD 550)	*Yu Pian* (《玉篇》), a book that arranges Chinese characters in the sequence of radicals	16,917	506
Ming Dynasty (AD 1368–1644)	*Zi Hui* (《字汇》), a vocabulary book whose arrangement of radicals was quite similar to that in modern dictionaries	33,179	573
Qing Dynasty (AD 1663–1772)	*Kang Xi Dictionary* (《康熙字典》)	47,035	960

Table 2.18. Trends of use of bamboo radical Chinese characters in various fields of society (Wu Danren, 1999).

Source	Years/Dynasty	Bamboo radical characters	Bamboo radical characters in specific fields (number/character)							
			Bamboo species and organ	Music	Culture	Agriculture and daily life	Transportation	Fisheries	Military affairs	Other
Oracle bone inscriptions (甲骨文)	Shang Dynasty (c. 16–11th century BC)	6	0	0	0	4	0	0	2	0
Inscriptions on bronze objects (金文)	Zhou Dynasty (c. 11th–3rd century BC)	18	5	2	1	7	0	0	2	1
Shuo Wen JieZi (说文解字)	Eastern Han Dynasty (AD 121)	151	20	20	13	82	4	1	2	9
Yu Pian (《玉篇》)	Liang Dynasty (AD 550)	506	164	37	14	223	11	8	5	44
Zi Hui (《字汇》)	Ming Dynasty (AD 1368–1644)	573	132	34	19	275	22	20	5	66
Kang Xi Dictionary (《康熙字典》)	Qing Dynasty (AD 1663–1772)	960	271	63	56	282	47	29	20	192

The poem highly praised bamboo's tenacity and courageousness in the face of severe conditions, fighting its way upward without fear.

President Xi Jinping, the current Chinese President, spent a tough countryside life in the rural areas of Shaanxi Province[1] for 7 years from 1969 to 1975, living in a cave house, sleeping on a *Tu Kang* (heated sleeping mud platform), eating bread made from corn, building dams, carrying manure and constructing methane tanks. In those difficult years, he revised the original poem of *Bamboo and Stone* by Zheng Banqiao to: 'Deeply rooted in the grass-root levels and does not relax, establishes firm stands among the people. Assailed by tens of thousands of hits, stand still no matter how the wind blows (深入基层不放松，立根原在群众中。千磨万击还坚劲，任尔东西南北风。).' The meaning here is 'Going through many hardships, still be strong and rightful.' This poem expressed Xi Jinping's ambition to embed himself at the grass-roots level, working hard together with people without fear of difficulties, and even devoting himself to help people. The poem also showed the appreciation of Mr. Xi of the bamboo spirits expressed by Zheng Banqiao in his poem.

In short, chanting for bamboos is the shining star in the art of Chinese poetry. These verses, with their natural, refined and elegant style, have offered people joyful feelings and inspired their lives.

In 2012, Jiang Zehui, and Peng Zhenhua, compiled and published the book *The Charm of Bamboo* (Jiang Zehui and Peng Zhenhua, 2012). This collects 100 selected poems and famous paintings about bamboo from the Ancient Qin and the Spring and Autumn Period (770 BC) to the late Qing Dynasty (AD 1854), as well as plenty of masterpieces related to China's bamboo culture.

Painting

Bamboo paintings have always been an important part of Chinese traditional painting. On the one hand, people appreciate them from the view of aesthetics. On the other hand, they praise them for their noble style, which had inspired the artists' creative thinking. The ancient Chinese people called paintings of bamboo 'freehand brush work of bamboo'. Bamboo was a popular object of paintings in ancient China. Bamboo paintings first appeared in the Han Dynasty and were already popular by the Tang Dynasty, during which period even murals of pitch-dark bamboo emerged. Up to the Northern Song Dynasty (AD 960–1127), Masters of bamboo paintings such as Wen Tong and Su Dongpo appeared. The famous pitch-dark bamboo artist Wen Tong was one of the originators of the famous Huzhou Bamboo School. Su Dongpo described his experiences of painting bamboos as: 'To paint bamboo, one must have a very detailed picture of a bamboo in mind.' This comment has been widely used today by Chinese people to describe a person who is well prepared and has a well-thought-out plan for missions to be carried out. The Song Dynasty Emperor Zhao Ji enjoyed painting bamboo so much that he placed bamboos as the background of his famous work *Listening to the Guzheng* (*Ting Qin Tu*, 听琴图).

Bamboo painting reached a higher peak in the Yuan Dynasty and the Ming Dynasty. A number of outstanding bamboo painting Masters emerged; these included Ke Jiusi (柯九思) and Li Yan (李衎) in the Yuan Dynasty, Wen Zhengming (文征明) and Xu Wei (徐渭) in the Ming Dynasty and Zheng Banqiao in the Qing Dynasty. Not only are their paintings of bamboo quite vivid, but the artists themselves also appreciated very much the spirits that bamboo symbolizes. What Zheng Banqiao expressed in his paintings were not merely the natural features of bamboo but also a wider range of connotations reflecting the features of his time and his spiritual essence. He loved bamboo deeply, and planted bamboos everywhere around his house, carefully observed them every day from morning to the evening, and produced many fabulous bamboo paintings. Nowadays, bamboo paintings are getting more and more popular.

Bamboo and music

Bamboo has a strong relationship with music, and people have special preferences for bamboo musical instruments (Jiang Zehui and Wang Wei, 2012). There are many bamboo music bands in Asia. In countries such as Japan, India and in the South-east Asian countries, bamboo musical instruments and bamboo music enjoy great popularity. In Latin America, bamboo musical instruments such as bamboo panpipes have been seen to be very popular. There are all kinds of bamboo instruments invented in the world

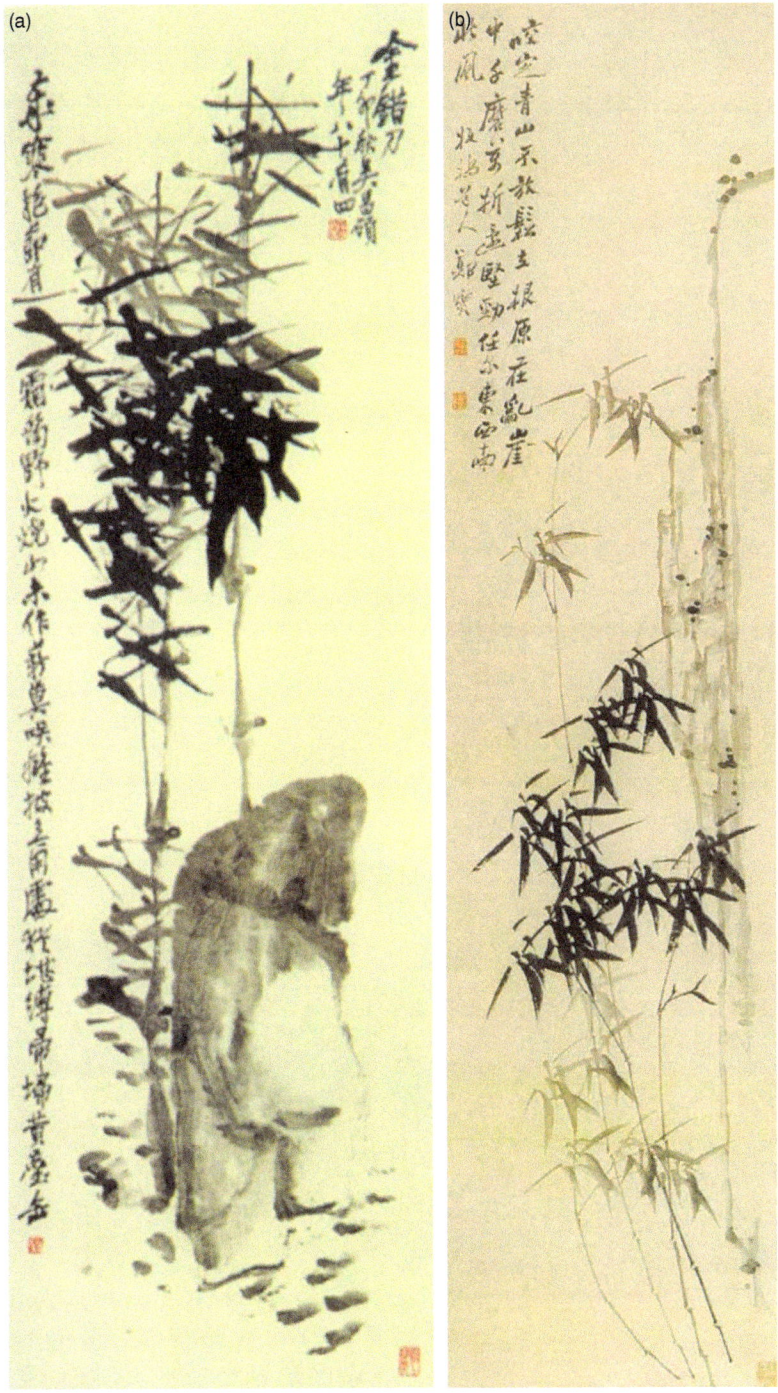

Fig. 2.90. (a) Painting by Wu Changshuo (AD 1844–1927). Available at: http://auction.artron.net/paimai-art5045261267/ (accessed 4 April 2017); (b) The painting 'Bamboo and Rocks' by Zheng Banqiao (1693–1765 AD). Available at: http://www.360doc.com/content/14/1212/09/16872596_432319883.shtml (accessed 4 April 2017).

Fig. 2.91. The painting '幽篁秀石图' by Gu An (顾安) of the Yuan Dynasty (reported in the link that follows as AD 1289–1365), and was one of the collections in the Palace Museum. Available at: http://www.hxsh365.com/newsShow.asp?dataID=26675 (accessed 28 June 2017).

and they all have their own specific features. Bamboo is one of the oldest materials for making musical instruments. The earliest bamboo instruments can be traced back to as long ago as 8000 years (Jiang Zehhui and Wang Wei, 2012). During the Zhou Dynasty (841–256 BC), bamboo musical instruments were classified in a separate category of instruments. At that time, musical instruments were classified into eight categories (or called the eight sounds or tones, 八音), namely: metal, stone, silk, bamboo, gourd, clay, leather and wood. Although we no longer use this classification,

we can see the importance of bamboo in ancient musical instruments.

According to the records of the Tang Dynasty, bamboo instruments accounted for half of all instruments, and the players of musical instruments were known as the bamboo people (竹人). The *Chinese Dictionary of Music* 《中国音乐词典》 (see http://baike.baidu.com/link?url= VZnxPIrAVvzZKP_wwZGYg9hC4-SDYNq-TO996teeXsyKhOSCGYfM204HIsPYFBbEr_Gsff-NduFTFObf34fo39NnnN4TbWhwTVY-9fnhoexl-H4D4_5UZ9KGIyezFCWittV88-JNZs1FsByFn-

Fig. 2.92. Painting by Yuan Yao (c. AD 17th–19th). Available at: http://www.missyuan.com/thread-101494-1-1.html (accessed 4 April 2017).

FMBJLG6_; accessed 10 July 2017) has recorded 90 cases of blowpipe musical instruments, among which 47 were made of bamboo, accounting for 52% of the total. The world's most popular musical instrument classification is based on the acoustic vibration of the instrument, which divides bamboo instruments into five categories: chordophones, aerophones, idiophones, membranophones and electrophones. Given the current wide range of bamboo instruments, they encompass all these five categories, even the modern electrophones. In China, 'sizhu' (silk and bamboo) are synonymous with orchestral music, with 'Si' (silk) representing stringed instruments, while 'Zhu' (bamboo) represents wind instruments. However, with the emergence of new instruments, some of the formerly excellent bamboo instruments were replaced by instruments made from (tree) wood and some even disappeared. Therefore, we are bound to strengthen the protection of bamboo musical instruments as a splendid cultural heritage. In China, some bamboo instruments, such as the Lu Sheng (which consists of multiple bamboo pipes, each with a free reed, which are fitted into a long blowing tube), the Yuping flute (a kind of transverse bamboo flute) and the Jinzhou Sheng (a mouth-blown free reed instrument consisting of vertical pipes) have already been included in the lists of intangible cultural heritage at national or provincial levels.

The Dongxiao (a kind of vertical bamboo flute) and the Taisho musical instruments of South Korea are also designated as cultural heritage (Moon Soontae, 2015). Many festivals are named after bamboo instruments, such as the Bamboo Organ Festival in the Philippines,

the Bamboo Music Festival in Sabah, Malaysia, the Panpipe Festival in Bucharest, Romania and the Lu Sheng Festival in Guizhou Province, China. The 'Waliha', played in a pizzicato style, is a national musical instrument of Madagascar, as are the Shakuhachi of Japan, and the Angk Lung of Indonesia. While protecting the tradition and relaying the cultural heritage of bamboo music, it is also important to note the necessity of innovation to adapt and meet the needs of the modern age, thereby letting the culture of bamboo music flourish like bamboo plants. The Chinese virtuoso Wang Wei has concentrated on researching the bamboo musical instruments of China and the world since the beginning of the 1990s, and in the face of many difficulties and under tough circumstances, has invented and developed a series of modern bamboo musical instruments, and created China's own 'green' bamboo orchestra – the Beijing Bamboo Orchestra.

Bamboo and dance

Bamboo plays an important part in Chinese dancing and singing. The most popular ancient dances – the Chinese dragon dance and the lion dance – use bamboo to form the frames of the 'dragon' and the 'lion', and use cloth and paper for taping. The ancient novel *Meng Liang Lu* (《梦梁录》) in the Southern Song Dynasty (AD 1127–1280), describes the dragon dance prevailing at the time, and the body of the 'dragon' could be made of 100–200 sections, or at least 20–30 sections. The lion dance enjoys a longer history. A vivid record about the

Fig. 2.93. A bamboo organ performance in Indonesia (Zhu Zhaohua).

Fig. 2.94. The Beijing Bamboo Orchestra performs at Washington National Mall in the USA (Wang Wei).

Fig. 2.95. Large rain (bamboo) hats dance (Zhu Zhaohua).

Fig. 2.96. The dragon dance. The dragon is made of bamboo (Chen Yunhua, 2012).

Fig. 2.97. International workshop participants join the Bamboo Pole Dance with local primary students to celebrate the Bamboo Culture Festival in Lin'an, Zhejiang Province, China (Zhu Zhaohua).

lion dance was made by the famous member of the literati Bai Juyi of the Tang Dynasty. What is more, the props of the popular folk dances imitating running horses and clams in the water are also made of bamboo frames and paper. In China, many important elements of folk singing and dancing, such as the flower fans, flower umbrellas, large rain (bamboo) hats and small (bamboo) baskets are all closely related to bamboo. There are many traditional folk dances of ethnic minorities that utilize bamboo props, such as the Lu Sheng dance of the Yi people, the Bobbin Drum dance of the Miao people and the Bamboo Pole Dance of the Li people, which is especially popular in Southeast Asian countries.

Kuai ban is a widely spread folk art in which a story is recited to the rhythm of bamboo clappers. The clappers are made of bamboo chips without nodes. Bamboo is also commonly used as a prop in Chinese acrobatics shows, such as pole climbing, walking, jumping and tumbling, etc., which are thrilling for the audience.

2.5.4 Bamboo crafting

Bamboo is an ideal material for decorating because of its graceful form and smooth surface. It is soft and full of elasticity, and can be easily split and carved. Bamboo handicrafts have a great variety in Asia, Latin America and Africa; they have special features in each country and are magnificent expressions. These handicrafts enjoy a special distinction and have a long-standing history in the rich realm of Chinese art. Chinese bamboo handicrafts have strong artistic vitality, and are always beloved by the people. Whether in a splendid palace or in a remote mountain village, bamboo handicrafts can be found everywhere. Among the different types of Chinese handicrafts, bamboo weaving, bamboo carving and bamboo Yu crafts are the three most important categories.

The splendid bamboo weaving craft

In all bamboo-producing areas, sooner or later, there appear exquisite and featured bamboo woven products. I have seen many wonderful bamboo woven products in India, Nepal, Bangladesh, Vietnam, the Philippines, Thailand and Ghana.

Chinese bamboo weaving production can be dated back to the Neolithic age, which was 7000 years ago. As human beings took up residence, they began to develop pottery to store foods and help to prevent starvation. They used bamboo baskets as the models for these pieces of pottery. This is why the potteries of that time have plenty of woven textiles on their surfaces. Chinese characters describing bamboo woven products such as different kinds of baskets (篮篓筐笼) appeared in an earlier time of China's history, indicating that the bamboo weaving crafts emerged in the earlier periods of the history. A colourful bamboo woven mat and a bamboo woven box unearthed from a tomb of the Chu State (740–223 BC) during the Zhou Dynasty in Jialing, Hubei Province, provided evidence that by then there was already a trend for bamboo weaving to develop into finer arts. During the Warring States period (475–221 BC), the rise of the iron-smelting industry had a profound impact on the upgrading of the tools used for bamboo weaving, and the art of bamboo weaving reached a new turning point. A beautifully painted bamboo woven cover unearthed from WangShan Tomb No. 1 in Jialing, Hubei Province in December 1965, was found to be made of very fine bamboo strips only 1 mm wide and 0.1 mm thick. The woven pattern is fine and beautiful and is still applied by artists nowadays in China.

In China's history, there were a large number of records about bamboo weaving groups, weaving techniques, Masters of bamboo weaving arts and their fine works. Ma Fujin of Dongyang, Zhejiang Province in the late Qing Dynasty, for example, made a pair of tower baskets for the imperial court, which took him 3½ years to complete and used more than 100 vats of paint. One of the decorative articles on the baskets imitates a real silver coin of the age. An old lady tried to pick it up while appreciating the work, only when she touched it, she found that it was made of bamboo. These types of basket are now collected by the Imperial Palace Museum in Beijing.

After the foundation of the New China (the People's Republic of China established in 1949), the Chinese government attached great importance to the heritage and development of bamboo weaving art, and the craftsmanship experienced a new leap. For example, the bamboo dimensional weaving crafts of Zhejiang Province have created and developed large amounts of woven products, such as bamboo cases, boxes, trays, fans, lanterns, screens, jars

Fig. 2.98. Bamboo woven craft of a Chinese dragon in China (Zhu Zhaohua).

and animal shaped toys and appliances such as duck baskets, goose trays, chicken jars, and peacock and panda toys. These products were exported in large amounts to Europe and America.

Equally good at dimensional bamboo weaving, Fujian Province's main bamboo products include bottles, baskets, pots, jars, boxes, cups, screens, curtains, furniture and pillows. Different from the style of Fujian Province, Sichuan's dimensional bamboo weaving is finer and more delicate, and the products include beautiful vases, tea sets, plates and so on.

Sichuan Province also produces porcelain items covered with woven bamboo, which function both as beautiful ornaments and practical

utensils. This craft also applies to pottery, glass, wood, or lacquer products. With bamboo-woven ware attached to the outside of these products, both their aesthetic value and the added value are greatly improved. Qingshen County in Sichuan is known as the 'home of bamboo weaving', and is where the Museum of Bamboo Weaving Crafts is located.

In southern China, the daily life of the ethnic minorities is closely related to bamboo; they live in bamboo houses, wear bamboo hats and bracelets, and use knife baskets, bamboo stools, tables, backpacks and pottery wine jars covered with bamboo woven ware. After the implementation of the reform and opening-up policy in China, bamboo woven products

Fig. 2.99. Bamboo woven lamps (Zhu Zhaohua).

Fig. 2.100. A bamboo woven elephant at the China Bamboo Museum in Anji, Zhejiang Province, China (Zhu Zhaohua).

have developed exuberantly as not only fine art works, but also daily utensils. In Xing Yi County of Guangdong Province, almost all of the local farmers are skilful in weaving bamboo baskets, and are able to produce various types of baskets. The local government and the export trading company collaborated in organizing them to produce baskets for international trade, adopted the model of 'company + farmer households', and more than 1000 farmer households participated in the production. In 2003, the bamboo basket export value of the County exceeded 1 billion CNY. Thus, the bamboo woven products became the main source of income of the local farmers.

Fig. 2.101a,b. Porcelain ware covered with woven bamboo (from Chen Yunhua, 2013, China Bamboo Weaving Art (中国竹编艺术－云华竹编精品); unpublished source).

Bamboo carving craft

Bamboo carving craft has its unique style, and has enjoyed a long history in China. Bamboo carving was primarily started by the use of bamboo slips (竹简, zhújiǎn) for inscribing. Before paper was invented, Chinese characters were carved out on pieces of bamboo – inscribed bamboo slips – and these became the carriers of characters and books. It is said that the philosopher Hui Shi (390–317 BC) brought five carriages of bamboo books with him every time he travelled to give lectures. When Dong Fang Shuo of the Western Han Dynasty wrote a 3000 word report to the emperor, the bamboo slips of the report were so heavy that it needed two strong men to carry them into the palace. With the development of the inscribed bamboo slip, there appeared special graphic art engraved on bamboo, as well as the art of carving on bamboo products. Bamboo carving art originated in the Spring and Autumn Warring States period in the 7th century BC. In a Western Han Dynasty tomb in Changsha, Hunan Province, people unearthed a bamboo spoon with a handle engraved with a dragon graphic; the textile was quite fine and it was considered to be the earliest bamboo carving craftwork found to date.

Bamboo carving art has been loved by the Chinese people since ancient times, and some of the carvings were even selected as contributions to the imperial palaces. Quite a number of bamboo carving treasures are now kept in the Treasure Gallery of the Palace Museum and the Taipei Palace Museum. Besides the traditional single blade and double blade methods, bamboo carving nowadays includes new techniques such as relief carving, multilayered carving on

the yellow parts of the bamboo culm, hollowed out carving, EDM (electrical discharge machining) engraving, etc. These techniques make the art of bamboo carving quite diversified.

The carved objects are mainly landscapes, people in stories, flowers, birds and animals. Ever since the mid-Ming Dynasty, bamboo carving art has become specialized, and at least 180 famous artists have been recorded in the ancient books. The two recorded famous styles of bamboo carving art – the Jiading Style and the Jinling Style – focused on artistic concepts and sentiment, and integrated poetry, calligraphy, painting and seal cutting in one; these works have a strong artistic appeal. During the Qing Dynasty, bamboo carving underwent greater development and won appreciation from the imperial emperors. During the reign of the Kangxi Emperor, there were three Masters of Arts in bamboo carving from Zhejiang Province, also known as 'the three brothers' – Feng Xilu, Feng Xijue and Feng Xizhang – who were called into the palace to specially create fine bamboo carving arts. The famous Chinese bamboo carving artworks such as *The Cabbage-Shaped Pen Holder*, *The Arhat* carved from bamboo root and the *Pen Holder with the Carved Cattle Herding* picture, which are now preserved in China's National Palace Museum, were all works from the three brothers.

The raw materials for bamboo carvings, as well as bamboo tubes and bamboo splits (including the green and yellow parts), also include bamboo roots. Because the base part of bamboo has thick walls and dense nodes, the roots are fibrous like human hair, and while carving it upside down, it is easy to make a vivid human face shape. Bamboo root carving is more

Fig. 2.102a,b,c. Bamboo carving crafts, China (Zhu Zhaohua).

prevalent in Zhejiang, Sichuan, Fujian provinces, and because these carving products can vividly manifest various facial expressions of human beings, they are popular at home and abroad, and have become popular tourism products around the country.

Bamboo carving is also very common in Latin America. Guadua, the giant bamboo that is widely distributed in the region, has thick walls and its rhizome resembles a crocodile. It is an ideal material for artistic creations. Different from China's delicate and exquisite craftworks, the Latin American bamboo carving art looks wild and vivid. In some Asian countries, there are also very fine bamboo carving craftworks. Many Masters of wood carving art lived in the famous wood carving centre in Baguio City in the Philippines, and because of the logging ban on precious tree species, in 2011, the Philippine

Bamboo Foundation encouraged them to try bamboo carving, in which they achieved great success. They found that the local bamboo species *B. blumeana* is a very good material for carving, and their primary works were very much welcomed and preferred by the local government and people.

Yu crafts and arts

Bamboo Yu crafts can be divided into two types, one is called XiaoYu (Small Yu), and the other is called DaYu (Big Yu). The basic techniques of the two types of craft are the same, but the raw materials are different. XiaoYu uses both large and small sized bamboos, such as Moso bamboo and *Ph. sulphurea* cv. [var.] *viridis* (syn. *Ph. viridis*); while DaYu uses only large sized bamboos such as Moso.

Fig. 2.103. Bamboo carving crafts made of bamboo base parts and roots, China (Zhu Zhaohua).

Bamboo Yu craft is one of the traditional Chinese bamboo crafts. It has a long established history: the craftworks are recognized by people with their artistic beauty, and favourable features such as lightweight, economical and practical, strong and durable. The bamboo Yu craftworks are especially favoured in South China – almost every family has bamboo Yu products. The bamboo Yu craftworks can not only be indispensable everyday household appliances, but also fine arts for collections and displays. The bamboo Yu technique uses a number of specialized and simple tools: the bamboo culms are connected firmly with each other in a strong structural form, the whole craftwork does not need any metal nails. The bamboo Yu products mainly include furniture, such as chairs, tables, stools, beds, cool plates, couches, children's strollers, bookshelves, arm-chairs, etc.,

and many other exquisite handicrafts, such as the micro-structures of pavilions, terraces, pagodas and open halls, etc.

The tools of bamboo Yu craft are simple: they include machetes, knives, digging shovels, scrapers, node polishing planers, torches, pin saws and trunk files, etc. The key technical point, or the secret of the bamboo Yu craft, is the determination of the size and shape of the Yu Kou – the encircling part that realizes the seamless joint of the horizontal and vertical bamboo parts. A skilled craftsman can calculate the length of the horizontal bamboo pole and the size of the Yu Kou according to the circumference of the vertical pole and the required number of angles of the final product; square-shaped products (with four angles) need to bend the horizontal pole at 90°, hexagon-shaped (six angles) products need a 60° bend

Fig. 2.104. Bamboo carving crafts made of the culms of *Bambusa blumeana* in the Philippines (Zhu Zhaohua).

and octagon-shaped (eight angles) products need a 45° bend. From ancient China to the present day, the folk artists of Yu technology have all been able to use very simple methods to accomplish the exact calculation. Because the products of China's bamboo Yu craft are of low cost and high quality, in recent years, a number of bamboo Yu Art Masters have been invited to other countries such as Indonesia and Ethiopia to train local artisans. They were very much welcomed by local people, because they made it possible for the poor to be able to sleep on comfortable and strong beds, as well as to use quite a number of pieces of elegant shaped and practical household furniture. In 2011, the craft of Bamboo Xiaoyu was included in the list of state-level intangible cultural heritage. Yiyang in Hunan Province is considered to be the demonstration site, where Wei Lumian and his team pass on the craft of Bamboo Xiaoyu to others (Wei Lumian and Wei Yu'an, 2016).

2.5.5 The history of Indian bamboo utilization and development

(This section is contributed by Dr Ramesh Chandra Chaturvedi, from India.)

The word bamboo comes from the Kannada term 'bambu', probably from 'mambu', which was introduced to English via the Malay 'samambu'.

Fig. 2.105. International training workshop on bamboo Yu furniture (Zhu Zhaohua).

Fig. 2.106. Bamboo Yu microstructure crafts, Hunan Province, China (Wei Lumian).

History and origin

No definite records are available to establish the antiquity, history and origin of bamboo craft in Assam. However, it can be safely assumed that the craft has been practised since the misty past and maybe since the very dawn of civilization. In the early period in Assam, bamboo was held in

Fig. 2.107a,b. Bamboo Yu furniture (Chen Yunhua, 2012).

Fig. 2.108. Bamboo weaving in India (INBAR).

special reverence and it is still forbidden to cut bamboo on 'auspicious days'. It is a general belief that bamboo possesses an auspicious character and is of religious significance.

An idea about the flourishing state of cane and bamboo products of Assam even during the time of Bhaskara Varman (in the early part of AD 7th century), the king of Assam, may be gained from the following extract from *The History of Civilization of the People of Assam to the Twelfth Century A.D.* by Dr P.C. Choudhury (Choudhury, 1959).

Early literature refers to the well-decorated and coloured 'sital patis' (cool mats) used by the rich people. Mats were usually made of cane. The classical writers testify the abundance of cane in the forests of Assam. Ptolemy (A.D. 90–168), for instance, states that to the east of Serica, which we have identified with Assam, there were hills and marshes where canes were grown and used as bridges. Evidence of the production of other cane articles is also supplied by the 'Harshacharita' [(*The Deeds of Harsha*), a history of the Indian emperor Harsha by Banabhatta], which mentions stools of cane. The cultivation of bamboo and its use for various purposes are well known. Bana (7th century A.D.) again testifies to this highly developed craft. He states that Bhaskara sent to Harsha 'baskets of variously coloured reeds', 'thick bamboo tubes' and various birds in 'bamboo cages'. All these prove that various industrial arts were developed in Assam at an early period and were continued to be practiced till recent times, based on the traditions like those of the craftsmen of other parts of India.

KING UPARICHARA VASU AND THE FESTIVAL OF BAMBOO POLE. Uparichara Vasu was a king of Chedi in the Puru Dynasty (about 3000 BC, before the time of the Sanskrit epic the *Mahabharata*, which is believed to be dated at around the 4th century BC) He was known as the friend of the deity Indra. During his reign, the Chedi kingdom had a good economic system and much mineral wealth. This made a lot of merchants around the world come to the kingdom, which was also abundant in animals and corn. There were many towns and cities in the kingdom. Uparichara Vasu had a very special chariot, and introduced a festival in the honour of Indra. The festival involved the planting of a bamboo pole every year, in honour of Indra, after which the king would pray for the expansion of his cities and kingdom. After erecting the pole, people decked it with golden cloth and scent, and garlands and various ornaments. The story behind this festival and Chedi presented in the extract below from the *Mahabharata*, which describes how Indra (the slayer of Vritra, a serpent or dragon, which represents drought) blessed King Uparichara:

> The cities and towns of this region are all devoted to virtue; the people are honest and content; they never lie even in jest. Sons never divide their wealth with their fathers and are ever mindful of the welfare of their parents. Lean cattle are never yoked to the plough or the

cart or engaged in carrying merchandise; on the other hand, they are well-fed and fattened. In Chedi the four orders are always engaged in their respective vocations. Let nothing be unknown to thee that happen in the three worlds. I shall give thee a crystal car such as the celestials alone are capable of carrying the car through mid-air. Thou alone, of all mortals on earth, riding on that best of cars, shall course through mid-air like a celestial endowed with a physical frame. I shall also give thee a triumphal garland of unfading lotuses, with which on, in battle, thou shall not be wounded by weapons. And, O king, this blessed and incomparable garland, widely known on earth as Indra's garland, shall be thy distinctive badge. The slayer of Vritra (Indra) also gave the king, for his gratification, a bamboo pole for protecting the honest and the peaceful. After the expiry of a year, the king planted it in the ground for the purpose of worshipping the giver thereof, viz. Sakra. From that time forth, O monarch, all kings, following Vasu's example, began to plant a pole for the celebration of Indra's worship. After erecting the pole they decked it with golden cloth and scents and garlands and various ornaments. And the god Vasava is worshipped in due form with such garlands and ornaments. And the god, for the gratification of the illustrious Vasu, assuming the form of a swan, came himself to accept the worship thus offered. And the god, beholding the auspicious worship thus made by Vasu, that first of monarchs, was delighted, and said unto him, 'those men, and kings also, who will worship me and joyously observe this

Fig. 2.109a,b. A boathouse made of bamboo (KONBAC, 2013).

festival of mine like the king of Chedi, shall have glory and victory for their countries and kingdom. Their cities also shall expand and be ever in joy.

A bamboo staff, sometimes with one end sharpened, is used in the Tamil martial art of 'silambam', a word derived from a term meaning 'hill bamboo'. Staves used in the Indian martial art of gatka are also commonly made from bamboo, a material favoured for its light weight. The Dandia folk dance in Gujarat also uses sticks that could be of bamboo or (tree) wood to clap during the movements of the dance.

The impacts of bamboo on Indian people's lives

This section is based upon an account given by Sanjay V. Deshmukh *et al.* at the 5th International Bamboo Congress (Deshmukh *et al.*, 1998).

Bamboos have served the fishing communities in the coastal parts of west India (Maharashtra, Goa, Kerala and Karnataka) as an important part of their lives for ages. One of their deities, Kalkai, is named after a local bamboo species (*B. vulgaris*), which is used for making flags for the deities at the entrance to sacred groves, and different types of bamboo basket are associated with different events in life such as birth, marriage and death. Bamboo articles for supporting livelihood (e.g. devices for fishing nets, fish traps, grading and measuring and the transporting and storage of fish) are crafted within the families using them. Bamboos are also used in the construction of boats, for scaffolding, parking slots for canoes, walkways, props, fencing material, mats, baskets and pot holders.

In the north-east of India, particularly Assam, bamboo has been and still is in widespread use in the life of the people. There are quite a few villages with houses predominantly made of bamboo. A roof of bamboo mat below a corrugated iron sheet gives a good appearance, and provides insulation against heat and sound. Supporting columns and trusses in bamboo are also common. Full round culm, or half rounds (or splits), placed horizontally or vertically with cross stiffeners, form the walls. Bamboo mats are also used for the walls. Even the fences in the houses in the towns and cities are made of woven bamboo splits. Quite a few of the items (implements, tools and accessories) used in homes, farms and for fishing are typically crafted from bamboo by members of the family using simple tools, and traditionally, most families had the required skills. Items in common use in village homes include cots with a bamboo frame and grass rope weave, stools or chair equivalent, mats, curtains, hanging or floor-resting shelves, baskets of various shapes and sizes for grains, vegetable, fish and cooked food, and sieves.

The impact of bamboo on Indian culture

Bamboo has many influences on Indian culture, such as in music, dance and even religion. An example is the Veena Vanshi Yagya Mandap Yagya Stambh Bamboo dances in north-east India.

Fig. 2.110a,b. A bamboo pavilion (a) under construction and (b) after the completion of construction in Goa, India (KONBAC, 2013).

MUSIC. The do-tara is a common simple form of stringed instrument used at the time of annual festival (Bihu). The gogona is another instrument that is based on the variation of notes produced by striking two pieces of bamboo splits with each other. Dances with people jumping between bamboo poles that are moved apart and back to touch each other in a rhythmic fashion (very like the Chinese dances) are also typical in Assam and north-east India.

RELIGION. Bamboo and Hinduism have a very interesting connection. It is said that bamboo shoots should not be harvested by a Brahmin, as the harvesting is compared with killing a child of the family. Similarly, Brahmins should not plant bamboo or a banana – they usually hire a people of other castes to do so. However, at all Brahmin weddings, bamboo and a banana are mandatory.

In Kathmandu, bamboos should not be harvested on Sundays or Wednesdays, or on a new moon or a full moon night, while in eastern Nepal, bamboos are not harvested on Mondays.

It is believed that wherever Buddhism went, bamboos went along. Bamboos and Buddhism were intricately associated. As Buddhism spread,

it is believed that Hindus got scared and started attacking Buddhist beliefs, and because Buddhism was intricately tied to bamboo, they started by attacking bamboo. Perhaps that is why the traditional skilled craftsmen of Nepal are not the Brahmins but those of other caste groups.

All Hindu marriages are performed typically at the bride's place under a shelter (a Mandap) that has four bamboo poles at the four corners of a square space, embedded in the ground, followed by a square mesh at the top, made by several rows of horizontally placed bamboo poles tied to the vertical poles and woven using grass rope. The final covering is of thatch. The Mandap is erected at an auspicious day and time a few days before the marriage, and dismantled a few days after it, on separate ceremonies during which the elders of the family and village join to invite the deities of the community to be present at the Mandap and bless the couple and later thank them and bid adieu.

A dead body should always be carried on a bamboo stretcher, and the final KapalKriya (liberating the soul from the body) of crushing the head of the burnt body is also performed using a bamboo pole.

Fig. 2.111. Bamboo corrugated board, India (I.V. Ramanuja Rao).

(a)
(b)

Fig. 2.112a,b. Bamboo buildings in India (I.V. Ramanuja Rao).

Impact of bamboo on spiritual culture in India

In China, bamboo is respected as a noble person, and represents the highest and core ethics of the country's culture.

In India, there is no specific information available on the impact of bamboo on spiritual culture. The flute, which is made of a hollow bamboo shaft, is the favourite of Krishna, one of the most important and widely worshipped Hindu deities. He is known as the holder or player of the flute and is invariably depicted in portraits and statues with a flute at his lips. In sermons preaching spirituality, people are advised to purge their minds of all thoughts, like the hollow flute, so that the pure voice of the almighty may flow through them. The music from flute is always associated with devotion and a calm mind rather than the boisterous notes of percussion instruments that are associated with Shiva (who is also referred to as Natraj or the supreme dancer). The Veena is a stringed instrument that is comparable in its association with spirituality, as its music is soft, and so ideally suited to meditative devotion, and is associated with the goddess of learning, Saraswati.

2.6   The Specialities of Bamboo Use in Environmental Beautification, and in Landscaping and Gardening

Bamboo has been widely recognized as an important garden plant in many countries in the world. Many Asian countries, such as Japan, Thailand, Vietnam, the Philippines, Indonesia and India, etc. have long histories of using bamboos for gardening, landscaping and ornamental purposes. Bamboos are considered to be an important resource for landscaping and gardening in these countries. Many European and American countries are also increasingly using bamboos in gardening.

China's garden art has long been closely related with bamboos. Many scholars in the ancient dynasties expressed their love and appreciation of bamboos in their literature. China's Ancient Gardening started as early as in the Zhou Dynasty (1029–771 BC). In the ancient manuscript *Strategies of the Warring States* (Zhan GuoCe, 《戰國策》) by Liu Xiang (77–76 BC), there is a chapter called Le Dun Fu (戰國策 • 乐敦傅) in which there are records of transplanting bamboo from the Hanshui area in south China to areas in north China near Beijing. The story also describes Qin Shihuang (259–210 BC), introducing bamboos for his imperial garden. The Tang Dynasty was the heyday of Chinese feudal society, when China's gardening achieved great development. In that time, many scholars considered bamboos to be good companions, and constructed many bamboo-featured gardens. The most famous poet in China's history, Du Fu (AD 712–770), and member of the literati Bai Juyi (AD 772–846) were both very fond of bamboos and admired them as their lifelong friends. Respectively, they made statements such as 'In the habitats of my whole life, there must be planted bamboos (平生憩息地,必种数竿竹)' and 'to reside in a place, one must manage a garden, for a garden one must plant bamboos (居必营园,园必植竹)'. After the Tang Dynasty, when Chinese gardening achieved further and greater developments in the Song, Ming and Qing

Fig. 2.113. A street scene with bamboos, Indonesia (Zhu Zhaohua).

Dynasties, bamboos were utilized more widely in gardening, and in these the designs were more exquisite, and the scales of bamboo landscaping larger. Why have bamboos so long been favoured in Chinese gardening and landscape construction? The reason is very simple, which is that bamboos are beautiful and they are easy to plant, so they will quickly give the required landscape effects (Lou Chong, 2007).

2.6.1   Bamboo has a unique natural beauty

Bamboos, whether sympodial, monopodial or amphipodial, and whether in a colony, such as a forest or plantation or a single stand, always has a beautiful appearance. A large bamboo colony will form an incomparably beautiful landscape. People will never forget the most famous scene in the Oscar-winning movie *Crouching Tiger, Hidden Dragon* (《臥虎藏龍》), in which the leading actor and actress fly over bamboo tops, wielding their swords to fight with each other. Because the bamboos in the scene were high, straight and graceful, they looked especially charming when they were swinging from side to side – the audience was not only immersed in the heart-gripping story, but also overtaken by the stunning scenery.

Large areas of bamboo forests are known as 'bamboo seas'. When tourists are on the scene

Fig. 2.114. A scene from the movie *Crouching Tiger, Hidden Dragon* (臥虎藏龍), which was shot in Anji, Zhejiang Province, China. Available at: http://www.3dmgame.com/news/201110/34325_2.html (accessed 4 April 2017).

themselves, they will have an incomparably pleasant and comfortable feeling. There are several 'bamboo sea' attractions in China, such as the Anji Bamboo Sea in Zhejiang Province and the Shunan (South Sichuan) Bamboo Sea in Sichuan Province. They are both very popular scenery sites that attract large numbers of tourists.

In other countries and regions in the world, such as Colombia in Latin America, there are the very eye-catching Guadua bamboo forests. I have often had the honour to visit Colombia's world-famous coffee production centre, and I found that the place was surrounded by beautiful Guadua bamboo forests. The pleasant presence of the Guadua bamboos made it feel more enjoyable to drink fragrant coffee. In India, if you find a clump of the giant bamboo (*Gigantochloa*) around you, you will feel that everything is vibrant, and the scene will provide you with comfort and strength. I have also been to Tanzania, by Lake Victoria, where nearly every village and every household has planted bamboo (*B. vulgaris* cv. Vittata) around the houses. This is a beautiful bamboo with yellow and green coloured culms. Local people think that it cannot only be used for housing, fencing and woven products, but is also very beautiful and can be used as a landscape attraction.

The aesthetic features of bamboo are reflected in many aspects, such as their morphology, colours and dynamics, and the artistic conceptions of bamboos all reflect their particular natural beauty.

Beauty in morphology

A single stand of bamboo is round, giving the feeling of mellowness and fullness. The culms are hollow inside and have internodes, and the whole culm is a combination of vigour and suppleness. A bamboo grove or forest has rhizomes that connect all of the stands, which are of symmetrical sizes, and look neat and orderly. In a clump of bamboo, the stands embrace each other closely, their canopy forming a beautiful crown, with luxuriant foliage and spreading branches, while the tips droop down naturally. Because of their morphological characteristics, a scene of bamboos gives a feeling of natural beauty.

Beauty in bamboo colours

The dominant hue of bamboos is green, and it is an extremely soft emerald green. Bamboos are evergreen, giving visual comfort when people appreciate them, but they are not monotone. The shoots can be very colourful, and those of different species can present with almost all of the colours found in nature; they also give a vibrant, positive feeling. The culms also have different colours – cyan, yellow, red, purple and black. The culms of some bamboo species have very beautiful and elegant stripes or spots, and a few species of bamboo have square-shaped culms. The leaves of bamboo not only vary in size, but also in colour – they often have yellow and white stripes, which have a mesmerizing beauty.

Fig. 2.115. Bamboo forest, water and a fisherman (Yang Jusan).

Dynamic beauty

A bamboo forest may present very different scenes over a year, giving it a dynamic beauty. In a year, a bamboo forest usually experiences different growth phases such as shooting, growing, branching, leafing and maturing, etc. In China, most monopodial bamboos grow in the subtropical zone where there are four distinct seasons, and the above-mentioned phases usually occur and are completed by the spring. In the summer, the bamboo forests are lush, thus providing cool shade. In autumn's cool breeze, the bamboo's elegant shape is reflected on the ground by the clear and white moonlight, which brings a leisurely and carefree mood. Bamboos in the winter show their toughness and spirit of unyieldingness under severe conditions. The tropical sympodial bamboos present another type of dynamic beauty. Over a year, most of the time there are new bamboos arising abruptly from underground, and the joining of the new generation to the old generations represents the virtue of harmony and common prosperity in a big family of several generations.

Artistic conceptual beauty

'Yi Jing' (artistic conception) defines the sensibilities, imaginations and resonations in the mind inspired by the form, connotation and charm of an aesthetic object. It is the uniform of the 'Real World' and the 'Virtual World', and thus, it is what we usually describe as the 'sensibilities or familiar feelings or emotions triggered by the scene'. The Qing Dynasty poet and artist Zheng Banqiao described the pleasant feelings brought on by the bamboo stands at his backyard as follows:

> A small and modest library, with a patio, several stands of elegant bamboo, and a few decorative stones, these do not take much lands, nor much costs. I could hear the bamboo in wind and rain, I could appreciate their shades under the lights of sun and moon, I drink and make poems with them, they are my companions in leisure and loneliness. Not only do I love the bamboos and stones, but they love me also.... In one room, I could see ten different scenes, though we are together for a long time, such sensibilities and tastes are unfading in my mind!

This is the rich imagination of a member of the literati who lived with bamboo, and the bamboo brought him great pleasure. What we have mentioned above, concerning the spirits and personalities of bamboo, are all artistic conceptual beauties of bamboo.

2.6.2 Advantages of bamboo in landscaping and gardening

In addition to the aesthetic features of bamboo that have been described above, the following

Fig. 2.116. Rural home inns in bamboo forests, Lin'an, Zhejiang Province, China (Zhu Zhaohua).

characteristics also make bamboo unique in landscaping and gardening.

The rapid formation of landscape effects

Generally speaking, when transplanting bamboos, the mother stock is used directly. Therefore, the landscape effects can be achieved immediately after transplantation. New shoots usually come out in the second year and grow into new bamboo stands. The size of all the stands in a grove or a plantation is comparatively unified. This is a feature that common tree species cannot match.

> **Once planted, bamboo can form a beautiful landscape in the same year.**

Strong adaptability and diversity

When developing landscapes and designing gardens with bamboo, there are always many different models to choose from, which are adapted for different climate and soil conditions while at the same time meeting people's needs. China has more than 500 species of bamboos, belonging to 37 genera. Among them, more than 100 can be used for landscaping and gardening, and there are plenty of choices for foliage purposes, stem purposes or shoot-producing plantations. There are also large, medium and micro-sized (grass-sized) bamboos that can be configured for use in comprehensive landscapes, or in small-scale gardens and scenes. Bamboos can be planted in large-scale plantations, or in clumps, or in coigns (projecting corners or angles of walls), or as bonsai. Bamboo can also be used as accessories to other gardening materials, such as other types of plants, decorative stones for hills and water bodies, as well as structures and buildings.

Economic advantages

Bamboo plantlets (in most cases bamboo mother stock) are generally of low cost, which is a great advantage compared with trees. Also, once planted, bamboo can generate income every year, providing steady production of shoots, culms or mother stock. Nevertheless, some special ornamental bamboo species, such as *Ph. heterocycla*, *B. ventricosa* and *Ph. nigra*, can also have quite expensive culms or mother stock. However, the use of bamboo as the main component for landscaping and gardening is absolutely viable from both the landscaping and the economic benefit aspects, as it can help to increase revenue and reduce management inputs.

Fig. 2.117. The appearance of a landscape a year after bamboos were planted (Zhu Zhaohua).

Fig. 2.118. Bamboo for gardening, Lin'an, Zhejiang Province, China (Zhu Zhaohua).

Fig. 2.119. Bamboo for the beautification of house fences, Hanoi, Vietnam (Zhu Zhaohua).

Ecological advantages

Bamboo forests have strong water and soil conservation capacities; they can also play roles such as air purification, without any contamination by pollen, and carbon sequestration at the same time. Large-scale bamboo landscaping forests can be an organic part of various ecological projects, including soil and water conservation projects, mining area reclamation projects, carbon sink forests and economic forest projects, and they may bring synthetic (improving) effects to the project area involved.

Fig. 2.120. Bamboo in a yard, Jiangsu Province, China (Zhu Zhaohua).

Fig. 2.121. Bamboo in the community, Jiangsu Province, China (Zhu Zhaohua).

Fig. 2.122. A beautiful environment surrounded by bamboo forest (*Arundinaria alpina*) in Africa (Jayaraman Durai).

Fig. 2.123. A bamboo raft and boat in Africa (Jayaraman Durai).

Fig. 2.124. A bamboo boat in Africa (Jayaraman Durai).

2.7   Summary: Bamboos and the Sustainable Development Goals (SDGs)

In this chapter, we have briefly introduced the characteristics of bamboos, and the contributions that they make towards the material and spiritual aspects of civilization and the living environment of human beings, as expressed by the title of the chapter, 'The Contribution of Bamboo to Human Beings Is Far More Than Is Imagined'. The descriptions that have been given of the characteristics of bamboos express messages that accord with those in the INBAR Position Paper posted on 24 August 2015 'Bamboo, Rattan and the SDGs', which was prepared for the UN Summit for the Adoption of the Post-2015 Development Agenda (25–27 September 2015). The paper (INBAR, 2015) started as follows:

How countries can harness these resources to add value to action plans for sustainable development

Six of the 17 Sustainable Development Goals (SDGs) to be debated and adopted at the UN General Assembly in September are directly relevant for bamboo and rattan producing countries and their green economy plans. They target: poverty reduction; energy; housing and urban development; sustainable production and consumption; climate change and land degradation. They all contribute to a seventh Goal – stronger implementation and partnerships. INBAR Member States and other producer countries can use bamboo and rattan to improve their national plans and add value to the global sustainable development agenda. Bamboo and rattan can also make a positive contribution to other SDGs addressing food security, women's empowerment, economic growth and technology.

Principally, the authors agree with the paper's statement on the contributions of bamboos to the SDGs. In the paper, seven SDGs were recognized to which bamboos could make actual and important contributions.
These are:

Goal 1. End poverty in all its forms everywhere.
Goal 7. Ensure access to affordable, reliable, sustainable and modern energy for all.

Goal 11. Make cities and human settlements inclusive, safe, resilient and sustainable.

Goal 12. Ensure sustainable consumption and production patterns.

Goal 13. Take urgent action to combat climate change and its impacts.

Goal 15. Protect, restore and promote sustainable use of terrestrial ecosystems, sustainably manage forests, combat desertification, and halt and reverse land degradation and halt biodiversity loss.

Goal 17. Strengthen the means of implementation and revitalize the global partnership for sustainable development.

At the same time though, we strongly support the inclusion of Goal 2, Goal 3, Goal 5 and Goal 8:

Goal 2. End hunger, achieve food security and improved nutrition and promote sustainable agriculture.

Goal 3. Ensure healthy lives and promote well-being for all at all ages.

Goal 5. Achieve gender equality and empower all women and girls.

Goal 8. Promote sustained, inclusive and sustainable economic growth, full and productive employment and decent work for all.

Why do we support the above four goals? Because Chapter 2 has provided sufficient information and facts on the contribution of bamboos to these four additional SDGs. For example, the importance of bamboos to Goals 5 and 8 is no less than its importance to the seven goals that were mentioned in the INBAR Position Paper.

The most essential issue is that all bamboo-producing countries, while developing their own bamboo sector, need to adopt the concepts and principles of sustainable development, and a suitable development model under the framework of sustainability. These are exactly the issues that Chapters 3 and 4 of the book discuss – we are trying to answer the question of 'How to develop bamboo sustainably' through analysing the experience that has already been accumulated on successful and failed cases.

Note

[1] Liangjiahe Production Group, Wenanyi Commune, Yanchuan County, Yan'an Prefecture City, Shaanxi Province, China (陕西延安延川县文安驿公社梁家河大队).

References

Some of the references included here have only minimum details of the text concerned, and may be informal papers that have been produced for training courses, etc., but they are included in order to give some indication of the sources used for this chapter.

Barathi, N. (2013) *Bamboo to Energy*. Growmore Bio-tech Ltd, Hosur, Tamil Nadu, India.

Chen Yunhua (2012) Benevolence to the whole area, he used bamboo slices to pull the local economy (他用竹丝传播大爱 拉动一方经济发展). In: Chen Yunhua *Love is Full of the World* (《爱满天下》), p. 51.

Choudhury, P.C. (1959) *The History of Civilization of the People of Assam to the Twelfth Century A.D.* Department of Historical and Antiquarian Studies in Assan, Gauhati, India.

Clark, L.G., Londoño, X. and Ruiz-Sanchez, E. (2015) Bamboo taxonomy and habitat. In: Liese, W. and Köhl, N. (eds) *Bamboo: The Plant and Its Uses*. Tropical Forestry 10, Springer, Cham, Switzerland, pp. 1–30.

Deshmukh, S.V., Karnik, P.C., Balakrishna Bhatkhande, Deshmane, A.B., Patil, A.K., Borkar, M.R. and Kedar Kulkarni (1998) Bamboo-dependent livelihood patterns of fishing communities: a case study of the west coast of India. In: Arun Kumar, Ramanuja Rao, I.V. and Sastry, C. (eds) *Proceedings of the Vth International Bamboo Congress and the VIth International Bamboo Workshop, San José, Costa Rica, 2–6 November 1998*. INBAR Proceedings No. 10, International Network for Bamboo and Rattan [now International Bamboo and Rattan Organization], Beijing, pp. 173–186.

Gutierrez-Céspedes, G.H. and Mendes-Araújo, M. (2015) Large scale plantations of bamboo for cellulose and biomass held in northeast of Brazil. Keynote PowerPoint presentation September 20th, 2015 under the Theme: Propagation, Plantations and Management. In: *10th World Bamboo Congress Proceedings, 17–22 September, 2015, Damyang, Korea*. World Bamboo, Plymouth, Massachusetts. Available at: http://www.worldbamboo.net/wbcx/Powerpoints/Propagation,%20Plantations,%20Forestry/PlantationsforPaper_Brazil_GutierrezAraujo_Keynote.ppt (accessed 21 June 2017).

INBAR (2015) *Bamboo, Rattan and the SDGs*. Position Paper, International Network for Bamboo and Rattan [now International Bamboo and Rattan Organization], Beijing. Available at: http://www.un.org/esa/forests/wp-content/uploads/2015/11/INBAR_input_AHEG2016.pdf (accessed 19 June 2017).

Jiang Zehui and Peng Zhenhua (2012) *The Charm of Bamboo*. China Foreign Languages Press, Beijing.

Jiang Zehui and Wang Wei (2012) *Bamboo Musical Instruments*. China Forestry Publishing House, Beijing, China.

Joshi, P.K. (2013) *Bamboo and Now the Climate Change*. TERI University, New Delhi.

Judziewicz, E.J., Clark, L.G., Londoño, X. and Stern, M.J. (1999) *American Bamboos*. Smithsonian Institution Press, Washington, DC.

Kigomo, B.N. (1988) *Distribution, Cultivation and Research Status of Bamboo in Eastern Africa*. Kenya Forestry Research Institute, Nairobi.

KONBAC (2013) Konkan Bamboo and Cane Development Centre (KONBAC). Presentation to: *Asia Regional Bamboo and Rattan Workshop, December 10–13, 2013, New Delhi*. Ministry of Environment and Forests, New Delhi and International Network for Bamboo and Rattan (INBAR), Beijing.

Lou Chong (2007) *Bamboo Landscape Gardens of Jiangsu Province* (江苏竹子园林). Scientific and Technical Documentation Press, Beijing, China.

Ma Naixun (2001) Biodiversity and resources exploitation of bamboo in China. In: Zhu Zhaohua (ed.) *Sustainable Development of the Bamboo and Rattan Sectors in Tropical China*. China Forestry Publishing House, Beijing, pp. 47–58.

Maria S. Voronstsova, Lynn G. Clark, John Dransfield, Rafaël Govaerts, William J. Baker (2016) World Check List of Bamboos and Rattans. International Network for Bamboo and Rattan (INBAR), International Center for Bamboo and Rattan (ICBR) of the China State Forestry Administration (SFA) and the Royal Botanic Garden of the Great Britain (KEW).

Moon Soontae (2015) Value of the humanities on bamboo. Keynote PowerPoint presentation under the Theme: Community and Economic Development. In: *10th World Bamboo Congress Proceedings, 17–22 September, 2015, Damyang, Korea*. Organizing Committee for Damyang World Bamboo Fair, Damyang, Korea and World Bamboo, Plymouth, Massachusetts. Available at: http://www.worldbamboo.net/wbcx/Powerpoints/Community%20and%20Economic%20Development/ValuesofHumanitiesonBambooKorea_Woo_Keynote.pdf (accessed 14 June 2017).

Ohrnberger, D. and Goerrings, J. (1985) *The Bamboos of the World: A Preliminary Study of the Names and Distribution of the Herbaceous and Woody Bamboos (*Bambusoideae *Nees v. Esenb.) Documented in Lists and Maps. Subfamily Bambusoideae: Its tribes and Genera*. International Book Distributors, Dehra Dun, India.

Ramanuja Rao, I.V. (2013) *Bamboo is Both a Grass and a Tree*. New Delhi.

Seethalakshmi, K.K. and Muktesh Kumar, M.S. (1998) *Bamboos of India: A Compendium*. Kerala Forest Research Institute, Peechi, India.

Voronstsova, M.S., Clark, L.G., Dransfield, J., Govaerts, R. and Baker, W.J. (2016) *World Check List of Bamboos and Rattans*. International Network for Bamboo and Rattan (INBAR), International Center for Bamboo and Rattan (ICBR) of the China State Forestry Administration (SFA) and the Royal Botanic Garden of the Great Britain (Kew).

Villegas, M. (2003) *New Bamboo Architecture and Design*. Villegas Editores, Bogota.

Wei Lumian and Wei Yu'an (2016) *Xiaoyu Bamboo Art in Yiyang, Hunan, China* (小郁-益阳小郁竹艺). China Central South University Press, Changsha, China.

Wood Research Institute of the China Forestry Academy (1982), *Wood Physical and Mechanical Properties of Main Tree Species In China* (中国主要树种木材物理力学性质), China Forestry Publishing House.

Wu Danren (1999) *Basic[s] of [the] Bamboo Industry*. Hunan Science and Technology Press, Changsha, China.

Wu Zhengyi and Peter Hamilton Raven (2006) *Flora of China*. The Missouri Botanical Garden Press and Science Press, St. Louis and Beijing, United States of America and China.

Yang Yuming and Sun Maosheng (2005) Biodiversity Characters of *Dendrocalamus giganteus* Munro. and *Dendrocalamus brandisii* (Munro) Kurz. (龙竹、勃氏甜龙竹发生生物学特性研究). Southwest Forestry University, Kunming, China.

Yi Tongpei (2008) *Chinese Bamboo Annals*. China Forestry Publishing House, Beijing.

Zhang Qisheng (1995) *Industrial Utilization of Bamboo in China*. China Forestry Publishing House, Beijing, China.

Zhang Qisheng, Jiang Shenxue and Tang Yongyu (2002) *Industrial Utilization on Bamboo*. Technical Report No. 26, International Network for Bamboo and Rattan [now International Bamboo and Rattan Organization], Beijing. Colour Max Publishers, Hong Kong.

Zhang Ying and Tong Lili (1997) Experimental studies on anti-aging effect of the leaf-extract of *Ph. nigra* var. *henonis*. *Journal of Zhejiang Agriculture and Forestry University* 16(4), 62–67.

Zhou Guomo and Shi Yongjun (2012) Report from a PowerPoint presentation on *Bamboo Forests and Carbon Trade* (竹林资源及其碳汇功能) made by Prof. Shi Yongjun in Chinese to participants in INBAR training courses in 2012.

Zhou Haochen (2017) China bamboo sector is progressing towards real economy – treasure the lucid waters and lush mountains, while at the same time, steadily realize the sustainable economy goals (走向实体经济的竹产业：要绿水青山，也要行稳致远). *China Economic Herald* 8 March, 2017, p. B4.

Zhu Shilin and Ma Naixun (1993) *Chinese Bamboo Plants Annals* (中国竹类植物图志). China Forestry Publishing House, Beijing.

Zhu Zhaohua (2005) *A Report on the Visit to Colombia*. Armenia, Colombia.

Zhu Zhaohua (2013) The cultivation and management technologies of Moso bamboo. Presented to: *Asia Regional Bamboo and Rattan Workshop, December 10–13, 2013*, New Delhi. Ministry of Environment and Forests, New Delhi and International Network for Bamboo and Rattan (INBAR), Beijing.

3

Key Issues Affecting the Sustainable Development of the Bamboo Sector

3.1 Resources Are the Foundations of Bamboo Industry Development

Many cases of failure were due to the lack of a genuine understanding of the local resources.

3.1.1 Resource inventory is the primary task of bamboo sector development

All processing industries, before they start, should have clear answers to three questions. Where do the raw materials come from? What are the requirements for raw materials? Is there enough supply? These questions seem rather simple, yet are almost always neglected. Many unsuccessful cases of industrial development are due to not properly answering these questions.

In the case of the bamboo sector, what should the contents of the resource survey be? They should include the following aspects: genetic resources; origin (natural or cultivated); distribution; and production.

Genetic resources of bamboos

We need to find out what bamboo species are there, what the characteristics of these species are, including whether: their sizes are large or small; their culms are straight or curved; their nodes are smooth or bumpy; their walls are thin or thick; their internodes are long or short; their rhizomes are monopodial, sympodial or amphipodial, etc. It is also important to understand the scale of the distribution of bamboo species – which species are in the majority and which are in the minority.

It is of utmost importance to be clear on these questions, because it will affect the product that will be made out of the bamboo. If the bamboo culms are straight, the nodes are smooth, the internodes are long and the culm walls are thin, the species could be a good material for producing woven handicrafts, and young-aged bamboo culms should be selected for producing woven products. In contrast, if the bamboo culms are large sized, with smooth nodes, and long and thick walls, the species could be the best material for producing laminated bamboo boards, and mature bamboo culms should be selected for the products. If the nodes are dense, and culm walls are thick, the species could be good for use as raw materials for bamboo carving and charcoal.

The origin of bamboo resources

Bamboo resources are generally divided into three types. The first is of natural origin, usually located in natural forests, and mixed with other trees and bamboo species. It is not intensively managed, and the age structure of the culms is

© Zhu Zhaohua and Jin Wei 2018. *Sustainable Bamboo Development*
(Zhu Zhaohua and Jin Wei)

Fig. 3.1. A bird's eye view of the Bamboo Sea in Anji, Zhejiang Province, China (Zhu Zhaohua).

not rational and the yield is relatively low. The second type of bamboo forest is also of natural origin, but after several years' management, it would look like a plantation. Anyone who does not understand the history of the area of bamboo will find it difficult to distinguish natural forests from plantations. If you go to Colombia or Ecuador, and see the beautiful and lush Guadua bamboo (*Guadua angustifolia*), you would think it is a plantation, but in fact in these two countries, more than 90% of Guadua bamboo is of natural origin. Thus, we can see that some natural bamboo stands, through management with simple technology processing, can easily be transformed into high-yielding bamboo forests. The third type of bamboo forest comes from cultivated plantations, to which are generally applied standardized density and management measures; they are usually monocultured and the ages of the culms are comparatively clearer than in unmanaged natural forests.

It would be rather difficult to utilize natural bamboo forests that have not been intensively managed for industrial processing. This is mainly because there might be different species in one forest and it is hard to tell the ages of the stands.

With rehabilitation and formation through the intensive management of natural bamboo forests though, these stands become more suitable for industrial processing.

The distribution of bamboo resources

The distribution of bamboo resources has an important relationship with future processing industries. A map of bamboo distribution should be produced according to survey results. Details of the condition of different bamboo species and their distribution areas should be provided in the survey report. A concentrated distribution of large areas of bamboo forests in one location could be advantageous for industrial utilization. For example, in 2000, we went to central and southern regions of Ecuador. We could see Guadua bamboos everywhere. However, they were all distributed in separated small areas, and it was hard to find a Guadua forest of tens of hectares or over a 100 ha, as they were mostly in small pieces. When conducting a bamboo resource survey, one should be especially careful with the issue that decentralized distribution will directly increase the costs of transportation of raw materials.

Fig. 3.2a–f. Various bamboo species with various morphological features (Zhu Zhaohua).

Fig. 3.3. *Dendrocalamus sinicus* – the king of the bamboo world. Its largest diameter at breast height (DBH) is over 30 cm and its height can reach 30 m (Hui Chaomao).

The production of bamboo resources

Evaluation of the potential output of bamboo resources should include the respective estimated production of different species and the estimated production of the accessible resources. Here, please pay attention not to mix the two different concepts, as the total amount of bamboo resources is not equal to the total amount that is available for industrial utilization. In many cases, limited by various conditions, the existing resources might not be immediately utilized. Such cases include: inconvenience in transportation, poor quality, the existing bamboo species not meeting the requirements of expected

Fig. 3.4. *Bambusa textilis* cv. Purpurascens in Colombia (Zhu Zhaohua).

Fig. 3.5. *Phyllostachys aureosulcata* (Zhu Zhaohua).

products, etc. Also, different products, production scales and processing techniques require different bamboo species and raw material properties.

If our main products are bamboo handicrafts, the consumption quantity is relatively low but the requirement for raw material quality is rather high. In the case of engineered processing on an industrial scale, a stable and sufficient supply of quality raw materials of the suitable bamboo species is one of the keys to success. Here are some examples from China.

For a product to achieve large-scale industrial production, it is required to have a stable supply of raw materials. In China, to establish a relatively small-scale bamboo laminated flooring factory, for instance – a factory whose annual production capacity is around $100,000$ m^2 – at least 300 ha of well-managed Moso bamboo (*Phyllostachys heterocycla* var. *pubescens*) forests are required. In the case of comparatively larger sized tropical sympodial bamboo species such as *G. angustifolia*, *Dendrocalamus asper* or *D. giganteus*, the requirement is still 200–250 ha. For reconstituted bamboo lumber production, one production line (one presser), with an annual capacity of 4000 m^3, needs the workers to work two shifts a day, and consumes 20,000 tons of fresh bamboo raw materials every year, which is equivalent to 500–550 ha of suitable Moso bamboo forest, or 400–450 ha of large-sized tropical bamboos. In order to build a bamboo primary processing factory (for toothpicks, barbeque sticks, curtains, mats, bamboo carpets or other products) with a capacity of 1 ton/day, about 10 tons of bamboo raw materials are needed daily. Therefore, increasing the unit production and quality of the bamboo forests should be the most important priority of bamboo forest management.

Taking the example of Anji County in Zhejiang Province: in 1978, Anji's bamboo forest area was 43,200 ha, the standing culms were 95 million and the (annual) production capacity was 9.33 million culms; but in 2010, the total area saw an increase of 33% to 57,333 ha, with standing culms 170 million, an increase of 79%, while the annual production was 23 million culms, an increase of 147%. This is because in the 1970s and 1980s, 60% of Anji's bamboo forests were low yielding, so the government focused on the rehabilitation of these low-production bamboo forests, rather than on simply enlarging the plantation area. At the same time, a series of bamboo cultivation and management technologies were developed and a number of policies were implemented to support local bamboo farmers to intensify their management of the bamboo forests. As a result, the unit production of the local bamboo forests increased rapidly.

3.1.2 Identify strategies on the development and utilization of the main bamboo species

Necessity

Different bamboo species have their respective morphological features and properties and, therefore, have various strengths and weaknesses for processing into different products. There is no bamboo species that fits all types of products. The use of bamboos for different types of products needs to take advantage of their strengths and avoid their weaknesses. Hence, it is important to know the characteristics of each bamboo in order to come up with suitable utilization strategies.

Information needed for making strategies

While making development strategies for different bamboo species, in addition to considering their morphology, growth and physical properties, one should also understand their local utilization and distribution. It is important to know how the local people use these species and their opinions or suggestions for development and utilization. This local knowledge of and opinions on bamboos would be very valuable for the identification of a reasonable and practical strategy. The practical experience of the local communities may provide a view for a comprehensive, scaled-up strategy for local bamboo development.

For example, in 2000, I (herein after refer to the first Author – Prof. Zhu Zhaohua) visited Kenya with the INBAR/Chinese Bamboo and Rattan Expert Delegation. I understand that the main bamboo species distributed in the country is *Arundinaria alpina*, a bamboo species distributed in the highlands where the altitude is between 2290 and 3360 m, and the most widely distributed indigenous species in Africa. In the area around Nairobi, local communities use the bamboo for fences, supporting poles for nurseries, nursery beds or flowerings. I asked one of the local technicians if the shoots are edible, and the answer was that they never eat the shoots, not in the local rural communities. However, when I asked a local farmer, he told me that he used to see people in the mountains eat the shoots of this bamboo, and they like the food very much. Such information provided an important clue for the expert team to apply a better approach when identifying the strategy for use of this species – we then understood whether we should advise to aim to develop only culm utilization, or to develop for both culm use (in handicrafts, such as woven products, carvings and bamboo pole furniture, etc., construction, charcoal, and fibre uses such as pulp, textiles, industrial fibres, etc.) and shoot use (fresh shoots or processed shoots, such as boiled shoots, dried/baked shoots and fermented shoots) purposes.

Comprehensive utilization of the local species

The utilization strategy should not only consider utilization of the main parts, such as the culms and shoots, of the bamboo, but also consider utilization of the other parts, such as rhizomes, leaves, branches, sheaths, and the residues left after processing, etc. In the effort to fully utilize the biomass and other functions of the bamboos, a reasonable, scientific and comprehensive supply chain will gradually form, with a focus on one or two major products.

Market demands

The needs and demands of the domestic and international markets should also be included in the factors affecting the decision making on the final products to be made. The strategy should consider utilizing the special features of local bamboo species, to produce advantageous and special products meeting the needs of the target markets. The following accounts are of three exemplary species with wonderful features that have very high potential for development.

Qiongzhuea tumidissinoda is a bamboo distributed in the mountainous areas of eastern Yunnan Province and western Sichuan Province, China. The height of the bamboo may reach 6 m and the diameter is 1–3 cm. A distinguished feature of the species is the fantastic form of the stem and nodes (see Fig. 3.6). It is an ideal material for ornamental purposes, as well as wonderful for making walking sticks. The products of this species have been traditionally exported to the countries, and the earliest trade of the species can be traced back to the Han and Tang Dynasties in the history of China. The shoots are fleshy, crispy and tasty, and are usually processed into dried/baked shoots.

G. angustifolia is a large sized, thick and straight bamboo; the nodes are white coloured, the stands look very beautiful and, in my opinion, it is one of the most beautiful bamboos in the world. It is not only a fine species for ornamental purpose, but also the most important industrial and construction material in Latin America. The above-mentioned features may be familiar to most bamboo developers. However, this species has another special feature which is very appealing – the rhizomes of the bamboo are shaped like crocodiles, so they may provide ample opportunity for artists and craftsmen to make special furniture and craftwork from them.

Chusquea aff. *culeou* (*C. culeou*) is a bamboo species widely distributed in the south temperate areas of Latin America (Judziewicz *et al.*, 1999), it is not very tall, about 6–7 m, and 3–5 cm wide, and looks just like sugarcane. The special feature of the species is that it is solid. In the wintertime, the leaves are jade green, while the stems are purple. As it is an amphipodial species, the bamboo grows in dense groves. The stands of this species may not only be used for ornamental purposes, but can also provide fodder for ruminants in the wintertime when other fodder is lacking. The shoots of the species are very tasty, though this might not be known to most Latin Americans and is worthwhile to be introduced widely. It is likely that this species would be outstanding for bamboo fibre production.

It can be concluded from the above that *Q. tumidissinoda*, *G. angustifolia* and *C. culeou*

Fig. 3.6. *Qiongzhuea tumidissinoda* (Zhou Yuanjiang).

Fig. 3.7. *Guadua angustifolia* forests in Colombia (Zhu Zhaohua).

Fig. 3.8. *Chusquea culeou* in Chile (Zhu Zhaohua).

have their own respective specialties that could be utilized for developing unique products.

3.1.3   A case of a bamboo flooring enterprise in Ecuador to illustrate the importance of knowing resources

When Chinese bamboo experts first visited Ecuador, they discovered that there were sufficient bamboo resources to develop various types of bamboo products in the country. The main species found was *G. angustifolia*, which is a good timber-purpose bamboo species. In 1940, Ecuador introduced from Asia about 40 fine bamboo species (Yang Yuming, 2000). These species

include, but are not limited to: *D. asper*, a tall and straight bamboo which is good for both culm and shoot purposes; *D. semiscandens*, one of the best shoot-producing sympodial bamboos; *Bambusa lapidea* and *B. tulda*, medium-sized fine species for culm purposes – construction or furniture; *B. blumeana*, which is fine for culm, shoot and charcoal purposes; *B. intermedia*, a medium-sized bamboo good for culm and handicraft purposes; *B. vulgaris* cv. Vittata, a fine species for ornamental purposes and also good for pulp making; *B. ventricosa*, a good species for ornamentation, furniture and handicrafts; and *Phyllostachys aurea*, a medium and small-sized monopodial species good for both culm and shoot purposes, which could be used for poles, furniture

Fig. 3.9. The solid culms of *Chusquea culeou*, Chile (Zhu Zhaohua).

Fig. 3.10. Jayaraman Durai of India in *Arundinaria alpina* forest in Kenya (Jayaraman Durai).

and handicrafts. These bamboo species were growing very well in Ecuador, so the Chinese experts believed Ecuador had sufficient genetic resources for developing a bamboo industry with various products.

However, the bamboo resources in Ecuador have the following problems:

- Although Guadua bamboos are widely distributed and plantations were seen to be starting to develop, there were no large-scale bamboo forests; the forests are sparsely distributed and far from each other. The management was rather poor, and overharvesting had caused serious reductions of bamboo forest area, from 16,000 ha in 1985 to 6000 ha in 2000.

- Even though many fine species have been introduced into Ecuador since 1940 and have been growing well, only a few, such as *D. asper*, have been developed into certain scales of plantations. Most species were not extensively propagated and used.

Identify product development strategies according to local resources

In 2000, the Chinese expert group suggested that Ecuador should prioritize bamboo resource development. According to the recommendations of the expert group, on the one hand, industrially purposed bamboo plantations were to be established and developed; on the other hand,

the management and protection of the existing local bamboo resources was to be strengthened. From the industrial utilization aspect, considering that that the existing resources were not yet sufficient to support large-scale industrial processing, traditional applications such as bamboo construction, furniture, handicrafts and supporting poles for banana, etc. were to be considered for initializing development.

In order to diversify the bamboo resources in Ecuador, under the requirement of Grupo Wong, 50 bamboo species were introduced from China to Ecuador. Eventually, 38 of them survived and are now growing well in Ecuador. These 38 species include *D. sinicus*, known as the King of bamboos (see Fig. 3.3, and *D. brandisii* and *D. latiflorus*, two well-known species for shoot purposes (Zhu Zhaohua, 2000, 2005c).

In September 2002, when a second group of Chinese experts led by Prof. Zhu Zhaohua visited Ecuador (Zhu Zhaohua, 2002), they were surprised to find that a bamboo flooring factory – Tropical Hardboo – had been established and could

Fig. 3.11. A new plantation of *Dendrocalamus asper* in Ecuador (Zhu Zhaohua).

Fig. 3.12. Another new plantation of *Dendrocalamus asper* in Ecuador (Zhu Zhaohua).

produce beautiful floorings with the local bamboo species – *G. angustifolia*. However, the experts soon found that the available bamboo resources to the factory were not of sufficient quality and quantity to support industrial-scale bamboo flooring processing. The production of the factory was facing many problems:

- insufficient raw material supply;
- high transportation costs as the bamboo forests were decentralized;
- faulty design of the production line; and
- high energy consumption.

Besides the above problems, the machinery of the factory had more serious problems: it was mostly imported from Taiwan, China (henceforward 'Taiwan'), and designed for Moso bamboo processing, while the local Guadua bamboos are larger, harder and higher in moisture content than Moso. While the machines had been trialled for Guadua processing, they required a lot of debugging and adjustments. As such, when the factory used the equipment to process Guadua bamboo, there were many problems (see also Section 3.6.4, Case study 8).

The Chinese expert group concluded that at the time, Ecuador still lacked the adequate conditions to produce industrial scale products such as floorings. The factory lasted for less than a year, and then had to finally be closed. The advice from the Chinese experts at the time was to develop bamboo resources as a precondition, and to make plans on product structure and production scales based on the areas and actual production of the bamboo resource bases in different periods. They also suggested the proper bamboo species to be propagated and cultivated to meet industrial processing demands in the future.

It was a great excitement for the Chinese experts to notice later that the entrepreneurs in Ecuador had learnt the lessons from their past experience and had paid more attention to the development of bamboo resources for industrial purposes. By 2013, 2200 ha of *D. asper* plantations had been newly established and larger areas of Guadua forests were managed; these provided a sufficient base for industrial processing. Three bamboo panel processing factories were established by 2013,

Fig. 3.13. Bamboo beds in Ecuador (Zhu Zhaohua).

Fig. 3.14. Bamboo handicrafts in Ecuador (Zhu Zhaohua).

producing floorings, decorative panels, furniture and veneer. The success of the processing industry motivated entrepreneurs to invest more in bamboo plantation development. All in all, the bamboo industry of Ecuador entered a healthy development period, when bamboo resources and the processing industry were evolving simultaneously.

Fig. 3.15. A Guadua pavilion in Colombia (Simón Veléz, Colombia).

3.2 An Elaborate Bamboo Development Plan Is Key to Avoiding a Detour

3.2.1 Strategic planning is an important measure for avoiding major mistakes

The purpose of planning is to identify goals for a country, a region, a village or an enterprise to achieve in a specific period of time, the ways to achieve these goals and the measures and policies to guarantee the successful implementation of the missions. Planning should be made based on a clear understanding of the available resources – bamboo resources, human resources, as well as the financial, social and economic conditions, such as the market demands. The process of planning is to integrate all of these resources, to create the necessary conditions and lay a solid foundation for realizing the goals. A multi-participatory, well-elaborated plan may help the developer to avoid blind and unrealistic decision making, the violation of natural rules, as well as detours and use of the wrong technologies.

3.2.2 Identify the targets and contents of a strategic plan

The first step in strategic planning is to determine the target, such as whether the plan is made for a province, a county, a town or a village, or an enterprise. Identifying the target is equal to having a clear list of tasks.

Take China for example; The Development Plan of China Bamboo Industries is a responsibility of the China State Forestry Administration (SFA). While making the plan, the SFA will refer to reports from the major bamboo-producing provinces, listen to the advice and suggestions of experts from various related fields of work, and consult with major research institutes such as the Chinese Academy of Forestry (CAF), Nanjing Forestry University, Southwest Forestry University, Zhejiang Agriculture and Forestry University and the China Bamboo Industry Association. The National Bamboo Development Plan is a macroscopic guideline that sets the goals for the next 5–10 years, and provides the policies to guarantee the success of these goals.

Corresponding with the Plan, all bamboo-producing provinces will have their own respective plans to meet the national requirements and goals, while at the county level, the plans are going to be very specific, including:

- First, a summary of the present development status: evaluation of production quality; analysis of the present industrial scale and the development scope; analysis of total investments and outputs; evaluation of the effectiveness and implementation of related policies; progress of all the bamboo-producing towns; new research, technologies and products; technical services and training; evaluation of impacts on rural development; analysis of challenges and opportunities, etc.
- Second, a 5–10 year plan developed based on the above systematic review.

This 5–10 year plan will include:

1. Resource development and industrial development plans (for primary industry, secondary industry, tertiary industry).
2. Measures and policies to guarantee the goals are achieved, including policies on land user rights, taxes, credit loans, investments, harvest regulations, product transportation, marketing, etc. Inspections of past policies and their implementation status are reviewed and detailed annotations provided to the new policies.
3. Promotion of and training on new research results, technologies and products.
4. The incentive measures – reward institutions and individuals for achieving the best outputs and warn or punish those who achieve the least.
5. The construction of demonstration sites – 'demo' forests, products and enterprises.
6. Market development.

Plans at the community level are usually made with the participation of community members, township leaders and relative experts. Enterprise planning will be discussed in Chapter 4 (Section 4.4).

3.2.3   Making a plan should engage multiple stakeholders

Bamboo industry planning for the state, provinces, counties and villages should be guided by the government, while the government should also listen to suggestions from the experts, corporate bodies and farmers, and learn lessons from past experiences both at home and abroad. One part of the organization of the government should take the lead in the planning, with other relevant departments providing support. In China, the leading organization is usually the Forestry Department and the supporting organizations are from Agriculture, Finance, Taxation, Industry, Transportation, etc. Experts, major enterprises and communities will be invited to the draft discussion and to give their opinions. Once the draft gets approved by the government, it will be implemented in the target area. If the plan might have a great impact on the local socio-economy, it should also be submitted to the People's Congress of the appropriate level for deliberation and adoption. Once the People's Congress has adopted the plan, it will be legally protected; it will not be altered by any change of government(s).

3.2.4   Case study 1: the implementation of the plan (Lin'an, Zhejiang Province)

How did they overfulfil the plan?

Once the plan is made, the implementation of the plan becomes the primary mission. A series of measures should be taken to ensure the effective implementations. Here we will take the example of the *13th Five-year Plan* of Bamboo Industry Development in Lin'an County, Zhejiang Province, China. As one of the top ten 'Bamboo Hometowns of China' (Zhu Zhaohua, 2011a,b, 2012a), Lin'an is China's top bamboo shoot producer. With a total population of 510,000, a total area of 3126.8 km² and forest coverage of 74.9%, the 13th Five-year Plan of Bamboo Industry Development (from 2003 to 2007), under the *13th Five-year Plan* of the local forestry sector (Lin'an Forestry Bureau, 2003), specified the following goals:

- The total area of managed Moso bamboo forests be increased from 20,000 ha in 2002 to 23,300 ha, at a rate of increase of 16.5%; for the other local species, such as *Ph. nuda*, a local species for shoot purposes, an increase from 16,000 ha to 16,670 ha, a 4.2% increase; *Ph. praecox* and *Ph. vivax*, which are main fresh shoot producers, from 16,700 ha to 20,000 ha, a 19.8% increase.

- The number of standing Moso bamboos be increased from 5.2 million in 2002 to 10 million, a rate of increase of 92%;
- Annual fresh Moso bamboo shoot production be increased from 6700 tons to 10,000 tons, a rate of increase of 49.3%; the annual fresh shoot production of the other species be increased from 210,000 tons to 250,000 tons, a 19% increase.

- Total bamboo shoot production value (including the primary, secondary and tertiary industries) increased from 960 million CNY in 2002 to 1.8 billion CNY in 2007, a rate of increase of 87.5%.

As a result of the co-efforts of the stakeholders over the 5 years of the plan, the goals were overfulfilled. The total managed bamboo forest area was increased from 56,000 ha to 65,333 ha, exceeding the target figure by 9.56%.

Fig. 3.16. *Phyllostachys* spp. introduced to the Philippines from China (Zhu Zhaohua, 2012c)

The total production value of the bamboo shoot industry was increased from 960 million CNY to 2.49 billion CNY, exceeding the target figure by 38.3%. Total fresh shoot production increased to 280,000 tons, exceeding the target figure by 12%. These achievements indicate that the *Five-year Plan* was successfully implemented.

How could Lin'an have achieved these successes? They were closely related to the measures taken to implement the plan, which respected the development rules of the bamboo sector. These measures were:

1. An administrative office especially responsible for steering project implementation was established. The office is responsible for coordinating the actions of the various related sectors of the government, township, village levels, as well as households and enterprises – understanding their interests, needs and resolving any problems on the way.
2. Allocation of tasks (work division): the missions of the plan were divided among different government sectors, towns and villages.
3. Strengthened support was provided to the cultivation and management of bamboo resources. Priorities were given to the improvement of unit production and the quality of the stands. For example, financial and technical support was strengthened for the rehabilitation of low-yielding natural bamboo forests – rehabilitated Moso bamboo stands were subsidized at 750 CNY/ha and newly established Moso bamboo plantations were subsidized at 3000 CNY/ha. This financial support greatly facilitated the construction of high-yielding and high-quality bamboo production bases.
4. A favourable policy environment was harnessed for the processing enterprises. Leading enterprises developing high value-adding products or improved technologies were given various types of support, including land, loans, finance, taxation, water and power supplies, etc.
5. Great efforts were made to research on and extend new products and technologies, such as the introduction and extension of bamboo charcoal, bamboo vinegar, bamboo fibre products, and early shooting technology, etc.
6. The local bamboo society played a great role in linking and coordinating the various stakeholders, and fostering cooperation among communities, enterprises, and the sciences and technicians.

3.2.5 Adjustments, monitoring and evaluation of plan

In China, a strategic plan usually covers 5 years or 10 years, and this is also the case for a bamboo industry plan. However, during the implementation of the plans, some of the goals or other contents of the plan may be found not to be in accordance with the actual situation. These contents may be adjusted or changed through a series of procedures. This problem can be tackled by drawing up an annual implementation plan and the adoption of an annual monitoring and evaluation system, thus identifying problems that are found and resolving them in time. When making an annual evaluation, local authorities should reward the best performances and criticize or punish poor performances, and special recognition should be given to the households, enterprises and scientists who made the greatest contributions. This recognition should not only be honours, but also material rewards such as in cash or in-kind benefits.

3.2.6 Case study 2: China's bamboo industries development plan from 2013 to 2020

In June 2013, China's SFA released the 2013–2020 National Bamboo Industry Development Plan, which is chaired by the government and of great authority. First, the National Plan makes an assessment of the status of the bamboo industry so as to determine: the future development goals, including the goals and tasks for key provinces and counties; the measures and supportive system to achieve the objectives; key projects and key construction tasks; and the investment required. This plan describes the bamboo industry in the next 8 years. The plan has 79 pages and is the result of multi-party participation and careful drafting. In October 2012, China's SFA prepared the draft of the *Development Plan for China's Bamboo Industries (2013–2020)* (SFA, 2013), which was then circulated to 14 bamboo-producing provinces and 112 key

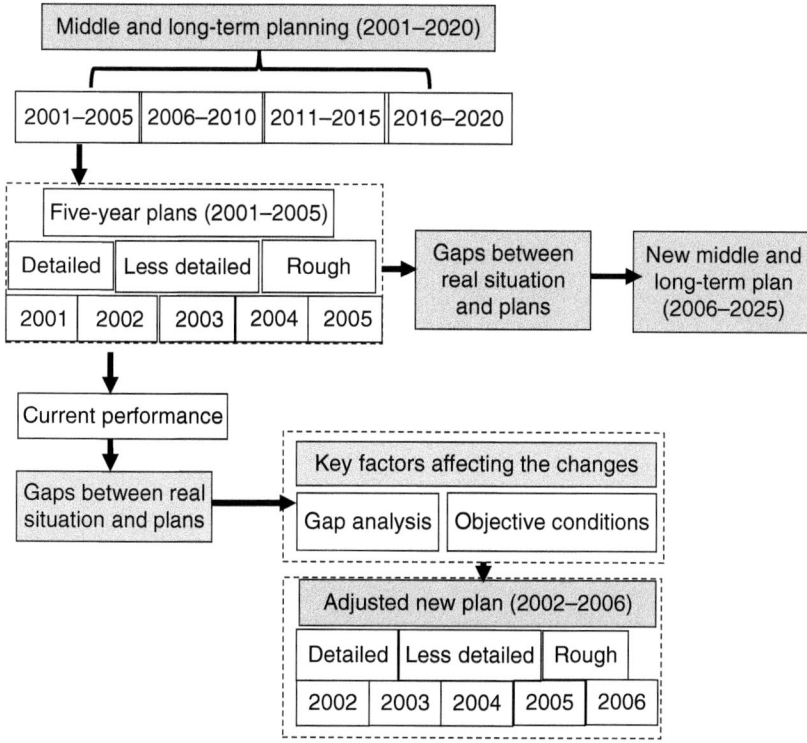

Fig. 3.17. Illustration of the middle and long-term plans of Lin'an County Government, Zheijiang Province, China (extracted and modified from Zhu Zhaohua, 2011b, p. 45).

counties, as well as experts and companies, to seek their views. During this period, a series of seminars and workshops was held for this plan. After 8 months of amendments and supplements, an official government plan was finally delivered. The 79-page plan is very rich in detail. An abstract of the plan is provided in Annex 1.

3.3 Product Development and Structure Optimization

3.3.1 Reasonable and effective product structure

Bamboo product development and orientation depends on the locally available bamboo resources, human resources and market demands. Product orientation should first focus on promoting and strengthening superior local products, and then gradually develop other new products. On the one hand, local superior products need to

be developed steadily, but on the other, various new products meeting market demands should be developed with a reasonable and full utilization of the available bamboo biodiversity and biomass, thus aiming to develop an effective product structure with high added-value.

Before investing in bamboo cultivation, forest management or processing industries, one should ask these questions:

- What products to be produced?
- Are there markets for the products?
- How is their profitability?

These questions are addressed in the rest of Section 3.3.

Make full use of different species and develop featured products

We have discussed in the earlier chapters the bamboo resource information that is needed for industrial development: this includes the

Fig. 3.18. Moso bamboo (*Phyllostachys heterocycla* var. *pubescens*) forest (Zhu Zhaohua).

Fig. 3.19. *Bambusa polymorpha* in Ecuador (Zhu Zhaohua).

available species, and their respective quantities/scales, distribution and quality, etc.

There are 1482 bamboo species worldwide and many of them have their unique features and can be used for different kinds of products. A considerable number of them could be used in scaled and industrial production. However, to date, there are only a limited number of bamboo species in the world that are utilized for industrial production.

Take China as an example; there are 600 to 700 bamboo species growing in China, but only around 50 of them are utilized for industrial processing, and over 80% of the bamboo culm production value and 50% of the shoot production value come from a single species – Moso bamboo. One reason for this is that people love Moso bamboo owing to the traditions and culture associated with it; another reason is that Moso bamboo has the perfect engineering properties that make it easy to be processed; it is also very good for ornamental and shoot/food purposes. What is more important is that the Chinese scientists have undertaken long-term, systematic researches on Moso bamboo and accumulated large amounts of research results that support the full development of this bamboo. In contrast, other bamboo species are comparatively much less developed because of a lack of technological knowledge. Therefore, there are still vast spaces for the exploration of bamboo utilization and development.

None the less, there are a few cases of good utilization of bamboo species other than Moso bamboo. In Guangning County, Guangdong Province (south subtropical China), there are many sympodial bamboo species that are of large, medium and small sizes, and they have been scientifically and reasonably developed for industrial uses. The large-sized species such as *B. pervariabilis*, *B. sinospinosa* and *B. beecheyana* var. *pubescens* are used for pulp and paper, while *D. latiflorus* and *Dendrocalamopsis oldhami* are used for producing shoots. The medium-sized species such as *B. textilis* and *B. chungii* are used mainly for producing incense sticks, and the sawdust can be used to produce briquette charcoal; because of their good splitting properties, which is appealing for woven products, and they are used to produce a special and elegant type of bamboo mat board furniture, decorative plates and woven decorative boards. Their young leaves are used for tea and beverages. The small-sized species such as *Pseudosasa amabilis* are used to produce various types of poles, and are exported in large amounts to Europe, especially the Netherlands, for use as fencing, decoration or handicrafts. For the local market, this species is widely used for producing furniture, curtain, fishing rod, etc.

We can see from the above that the major local bamboo species in this county are reasonably utilized for high value-added products. The product structure meets the diversified market demands. In 2013, the county's bamboo industry production value reached 1.3 billion CNY, which was the top of all sympodial bamboo-producing areas in China.

Make product plans based on local human resources and traditions

When setting up an industry, it is important to identify the local human resource conditions and traditions. Let us still take Ecuador as an example. Like the other Latin American countries, Ecuador has a strong cultural atmosphere, especially for Guadua culture. The people in Ecuador love bamboo, and love to stay in bamboo houses. The country has a long history of and advanced technologies for bamboo construction. There are exquisite bamboo houses for the rich, and simple but practical bamboo houses for the poor. Viviendas Hogar de Cristo, a church-based organization, has been running a bamboo housing factory, that built 70 bamboo houses a day (each with an area of 23–25 m²) for poor families, which consumed 100 tons of bamboo raw materials. Many handicrafts, tourism products and furniture in Ecuador are also made of bamboo. In 1999, the company FORESA was able to successfully produce bamboo plywood. All of these traditional and new industries, as well as the local experts, technicians and craftsmen, are a solid base for Ecuador to further promote its bamboo sector. In view of these conditions, the visiting Chinese expert group suggested a three-stage plan for the simultaneous development of bamboo resources and processing industries (Zhu Zhaohua, 2002).

THE BAMBOO PROCESSING INDUSTRY IN ECUADOR. The experts made assumptions on the future

Fig. 3.20. A plantation of *Pseudosasa amabilis* in Guangning, Guangdong Province, China (Jin Wei).

Fig. 3.21. A bamboo (*Pseudosasa amabilis*) pole processing factory in Guangning, Guangdong Province, China (Zhu Zhaohua).

Fig. 3.22. Bamboo (*Pseudosasa amabilis*) poles drying in Guangning, Guangdong Province, China (Zhu Zhaohua).

Fig. 3.23. Bamboo furniture made using the poles of *Pseudosasa amabilis*, Guangning, Guangdong Province, China (Zhu Zhaohua).

Fig. 3.24. A plantation of *Bambusa textilis* in Guangning, Guangdong Province, China (Wu Guangmin).

structure of the bamboo product industry in the short-, middle and long-term development plans of Ecuador. These were made on the basis of the decentralized distribution of production bases, large areas of mono-species forests – *G. angustifolia* – and small areas of distribution of other species, such as *D. asper*, *B. tulda* and *Ph. bambusoides*. The short-term plan was based on the present existing resources; the middle-term plan on an assumption of 10,000 ha of industrial plantation being established in 5–10 years; and the long-term plan on an expected plantation area of 50,000 ha with a reasonable composition of various fine species. The following is a list of suggested products to be made in the three periods:

1. Short term (1–4 years): the main products include bamboo houses, bamboo furniture, bamboo sticks (toothpicks, skewers), packing boxes (for vegetables, fruits, oil palm, etc.), bamboo handicrafts, shoot products, bamboo scaffolding, bamboo poles for supporting banana plants and bamboo woven products such as bamboo curtains.

2. Middle term (5–10 years): along with the short-term products, this plan aims at developing an annual bamboo panel production capacity of 200,000–300,000 m^2 (for decoration, cement moulding, roofing, etc.). The plan also included bamboo mats, bamboo charcoal and bamboo shoot processing industries.

Fig. 3.25. Jin Wei communicating with an old lady who was manually processing bamboo (*Bambusa chungii*) incense sticks in front of her house in Guangning, Guangdong Province, China (Zhu Zhaohua).

Fig. 3.26. Bamboo (*Bambusa chungii*) woven furniture and wall decoration panels in Guangning, Guangdong Province, China (Zhu Zhaohua).

Fig. 3.27. Plates made of pressed bamboo (*Bambusa chungii*) mat panels in Guangning, Guangdong Province, China (Zhu Zhaohua).

3. Long term (11–15 years): along with the short-term and middle-term products, this plan aims to produce 300,000–500,000 m² of bamboo panels per year, with a large industry of bamboo charcoal processing, as well as pulp making at a capacity of 50,000–100,000 tons/year.

Note that the above recommendations were made on a presumption that the bamboo resources in the country continues to develop in all three periods, but they lack any reference to market demands. However, all investments into new products need an overall study on the demands of domestic and international markets.

An expert estimation showed that if the development of the bamboo sector goes smoothly and successfully, within 15 years, Ecuador may have a bamboo industry with an annual production value of US$0.8–1.0 billion; this is going to make Ecuador an important country with some of the most comprehensive bamboo industries in the world.

3.3.2 Establish an extensive supply/value chain through the full utilization of bamboo biomass

Low utilization efficiency is the biggest challenge in the bamboo processing industry

In 1999, I was able to pay a visit to the Vietnam Forestry Institute. During the visit, I saw Mahjong mat products made of *D. barbatus*; they were of fine quality and are quite similar to Chinese products. I requested a visit the Mahjong mat factory but, to my disappointment, I was told that it had been closed. The reason was that a large amount of better quality, finely designed and cheaper Mahjong mats from China have taken over the domestic markets.

I started pondering on this problem: what were the real reasons for the failure of the Vietnamese enterprise? At that time, the raw material price and the labour wages in Vietnam were

Fig. 3.28. The bamboo processing factory Viviendas Hogar Cristo in Ecuador produces 70 sets a day of houses for the poor (Zhu Zhaohua)

respectively 30–40% and 40–50% of those in China. So what were the factors that led to the high price of the Vietnamese products? They may come from various aspects, such as production management, transportation, etc., but I believed the main reason was the low utilization rate of the raw materials in Vietnam. If the whole bamboo culm was only used for one type of product, and if the wasted materials were not utilized for other products, the utilization rate would be very low.

In June 2006, a Chinese expert group visited one of the bamboo factories of Vietnam on a consultancy project. The factory was producing composite floorings. The experts found that though the product qualities were fine, the raw material utilization rate was only 5–6%. Another factory nearby producing bamboo sticks only used 8% of the raw material. The large amount of 'wasted materials' from the factories piled up outside the factory, was hard to handle properly and became a source of environmental pollution – most of the waste was burnt or dumped into the

river (Zhu Zhaohua, 2006). As a result, many of the primary processing factories in Vietnam were located beside rivers. China's bamboo sector development used to experience this same problem. From the late 1980s to the early 1990s, the raw material utilization rate for bamboo mat and curtain factories in China was between 20 and 25%. During that period, each factory started from the raw bamboo culm and made the final products – there were no 'semi-product' suppliers, and so the low utilization rate of the raw culm caused serious waste problems.

The emergence of the primary processing factories, a revolutionary change in China's bamboo sector development

How did the 3000 types of bamboo products in nine categories develop in Anji County? – the magic of bamboo biomass utilization.

Fig. 3.29. A slum in Ecuador (Zhu Zhaohua).

Fig. 3.30a,b. Waste bamboo materials that have been burnt outside a bamboo processing factory in Vietnam in 2006 (Zhu Zhaohua).

While trying to cope with the high costs of raw material processing, some Chinese enterprises tended to offer primary processing machines to bamboo farmers, so that they could provide specific semi-products according to the requirements of factories dedicated to specific products, for example, provide flooring laths to flooring factories and strips to curtain/mat factories. The famers soon found that many parts of the bamboo culms were not utilized, such as the stem bases, top parts and branches, and that these materials might be valuable for other factories producing different products. So rather than supplying a unique product to one factory, the famers started to sell these unused parts of the bamboo culms to other factories that needed such (semi-)products. These early primary processing factories discovered that processing various semi-products for different

Fig. 3.31. Waste bamboo materials piled up along the riverside waiting for a flood to wash them away in Vietnam in 2008 (Zhu Zhaohua).

Fig. 3.32. Waste materials from a bamboo chopstick processing factory that had piled up and blocked the gateway in Hainan Province, China (Zhu Zhaohua).

factories could increase the utilization rate of the bamboo culms by turning the waste into valuable semi-products, and that they could profit from these operations.

By the mid and late 1990s, a large number of independent primary processing factories were established. These provided semi-finished products or raw materials to factories dedicated to specific final products. Take Anji for example; hundreds of primary processing factories emerged within several years and their average daily consumption of bamboo culms was 20–50 tons. The emergence of these independent primary processing factories brought revolutionary changes to China's bamboo sector, and was welcomed by all stakeholders:

* The manufacturers of specific products welcomed the changes. This change saved them a large amount of energy and also saved human resources in organizing the procurement and transportation of raw materials, as well as disposal of the unused parts; at the same time, they were able to access competitive and specifically made semi-products by selecting raw suppliers, which could greatly improve the quality of the final products. Estimates showed that the costs of raw materials could be reduced by 50% compared with the costs involved in processing the bamboo culms themselves.
* The past 'waste materials', which cost considerable resources to dispose of, have become products that are short in supply. For example, 40% of the raw material turned into sawdust by processing could be sold for 250 CNY/ton (2008).
* The utilization rate of raw bamboo materials grew from less than 25% to between 85 and 90%. The value added to the raw materials was dramatically improved – by two to three times, and so the price of bamboo culms has been increasing fast – from 160 CNY/ton in 1985 to 860 CNY/ton in 2010. This has resulted in a marked increase in bamboo farmers' incomes, and at the same time significantly reduced the environmental impacts of bamboo utilization.
* The costs of finished products also decreased (Zhu and Jin, 2006), and a healthy supply chain was formed in which the enterprises were relying on and independent of each other at the same time.

* The reasonable utilization of the raw material inspired the bamboo manufacturers to innovate and produce more varieties of products with better quality. With more and more new products appearing on the market, Anji's bamboo product structure became more and more reasonable. For example, nowadays, companies in Anji manufacture over 3000 different types of bamboo products in nine categories – bamboo panels, woven articles, shoots, handicrafts, fibre and textiles, biochemicals, biomass energy products, furniture and constructions. These products could meet the demands of quite a number of the diversified customers and markets.

Case study 1: a study on a bamboo primary processing factory in Anji

The different parts of the bamboo culm were utilized in the following way. When the culms arrived in the factory, they were first cut into four parts: the base part, which is good material for charcoal; the middle lower part, which is good material for flooring and furniture; the middle upper part, which is good material for mats, handicrafts, sticks, etc.; and the top part, which is good material for sticks, chopsticks, scaffolding and fencing (see Figs 3.33 and 3.34).

In 2005, we carried out an investigation on a primary processing factory in Anji; the investigation covered the portions and prices of the semi-products and wasted materials, the processing costs and profit. The findings are summarized below.

1. The factory's average daily consumption of bamboo culms was 25 tons. The diagram in Fig. 3.35 shows that if only the middle lower part and middle upper part were used to make semi-products, the utilization rate would be only 16%. During the finishing of the final products, there would be more waste, and the final utilization rate was estimated to be less than 14%. Thus, if the remaining 84% of the residues from processing were not reutilized, they would become wastes and be disposed of. In Fig. 3.35, 'Others' means dust and water lost through evaporation, and this part accounts for 8% of the total raw materials. Therefore, the possible highest utilization rate would be 92%; 100% is not possible (Zhu Zhaohua, 2012b).

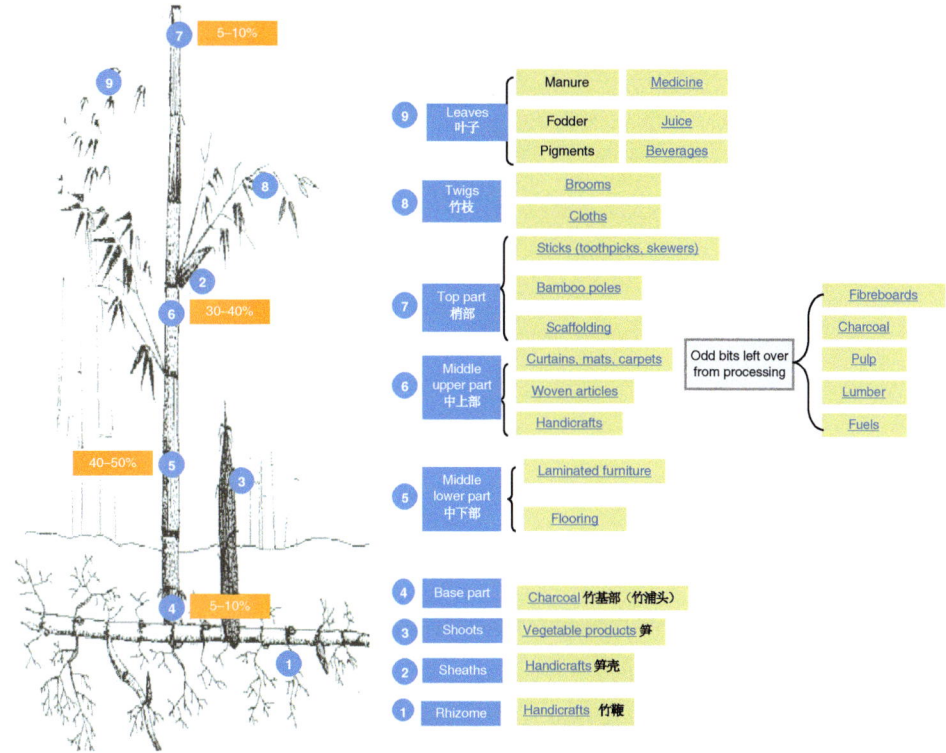

Fig. 3.33. Diagram of the utilization of the different parts of bamboo (Jin Wei).

Fig. 3.34. Raw cutting at a bamboo primary processing factory, Anji, Zhejiang Province, China, 2005 (Zhu Zhaohua).

2. The amounts and prices of different types of products and residues after primary processing were as follows (the asterisked figures in parentheses are figures that have been updated from 2005 to 2008).

Sawdust: 10 tons, 60 (240*) CNY/ton (40%).

Base part: 1.8–2.0 tons, 100 (200*) CNY/ton (8%).

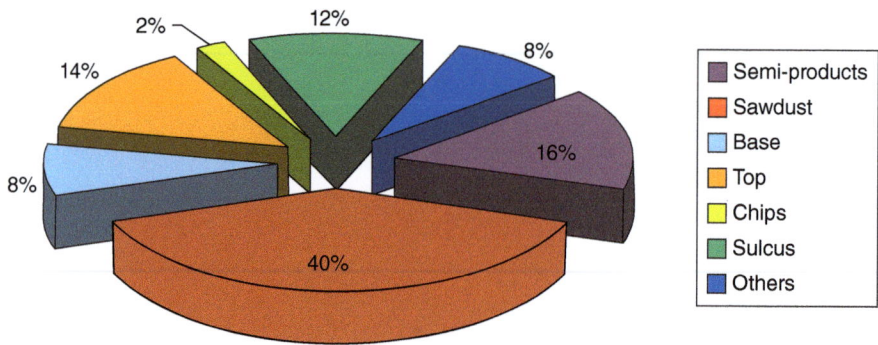

Fig. 3.35. Utilization rates of portions of different parts of the bamboo culm after primary processing. 'Others' means dusts and lost water through evaporation.

Top part: 3–4 tons, 300 (400*) CNY/ton (12–16%).

Wasted chips: 0.5 ton, 200 CNY/ton (2%).

Sulcus (or water way – a groove running the length of the internodes under the branches) portion:

3 tons, 350 (400*) CNY/ton (12%).

Others: evaporation and dust, 2 tons (8%).

3. The daily operation costs were as follows:

Raw material: 700 CNY/ton (25 tons = 17,500 CNY)

Labour: 1400 CNY

Insurance: 30 CNY/person/month, 200 CNY/day

Fuel: 2.5 tons/day

Depreciation: 66 CNY/day

Tax: 425 CNY/day

Profit: 2195 CNY/day

4. The value increase per ton after primary processing was 171.4 CNY, a rate of increase of 24.5%. (The bamboo farmer's income from branches, leaves, top parts and sheaths is about 2500 CNY/ha every 2 years; the income from shoots (winter, spring and rhizome shoots) is about 19,300 CNY every 2 years (this was the figure in year 2003).

5. Analysis of the daily profits of the factory in 2005 gave the following figures. The income from different products was:

Semi-products: 4 tons @ 4999 CNY = 19,996 CNY

Sawdust: 10 tons @ 60 CNY/ton = 600 CNY

Base: 1.9 tons @ 100 CNY/ton=190 CNY

Chips: 0.5 tons @ 200 CNY/ton = 100 CNY

Sulcus: 3 tons @ 350 CNY/ton = 1050 CNY

which gives a total income of 21,785 CNY and a profit of 2195 CNY/day.

The above statistics show that 91% profit of the primary processing factory came from the sales of residues (sawdust, base, chips and sulcus: 1940 CNY). At the same time, the finishing factories could save the costs of raw processing done by the primary processors, and could use this part of the (former) costs for improving product competitiveness and developing new markets. The cost-effectiveness of this supply chain will finally benefit the customers, who should get finely made products at comparatively cheaper prices.

3.3.3 Product structure optimization based on technology innovation and changes in market demands

With the emergence of new products and innovation of new technologies, customer requirements for bamboo products – variety, design and quality – are becoming higher and higher. Products that used to be popular will gradually be replaced, and all bamboo enterprises will be confronted with severe market competition. It is vital for these enterprises to continuously observe and analyse market changes, and be persistent in innovating new technologies and new products.

Changes in market demands and raw material prices provoke the optimization of product structure

Let us take China as an example. In the middle and late 1980s, China began to produce Mahjong

Fig. 3.36. Particle board factory, Anji, Zheijiang Province, China (Zhu Zhaohua).

Fig. 3.37a,b. Biofuel pellets, Anji, Zheijiang Province, China (Zhu Zhaohua).

mats, and when the products first entered the domestic markets, they immediately attracted a large number of consumers as they were beautiful and comfortable. Further, this type of product did not require complicated technologies or high investments. By the middle and late 1990s, when Mahjong mats were at their peak popularity, there were about a 1000 production lines in full operation in many of the bamboo producing areas of China. However, in the late 1990s and early 2000s, a new type of high-quality bamboo mat entered the market; in comparison with the new products, the weaknesses of Mahjong mats were that they were too heavy, and their manufacture produced too much waste of raw materials. By the middle 2000s, the production and market share of Mahjong mat products dropped dramatically to a very small scale.

Other products, such as laminated floorings and bamboo vinegar products, experienced

Fig. 3.38. Bamboo sawdust briquettes ready to be made into charcoal (Jin Wei).

Fig. 3.39. Large amounts of sawdust from the primary processing factory are automatically collected and loaded on to a truck, which transports the sawdust to other factories producing particle boards, biomass fuel pellets and charcoal (Zhu Zhaohua).

similar market changes. Below (see Case study 2), there is a review of the evolution of bamboo product structure in Anji, and of the associated research and new technologies from 1986 to the present. Besides the market demands, the other important factors affecting the product development and product structure evolution were: the supply and prices of raw materials, environmental impacts of the industries and the policy environments (Zhu Zhaohua and Chen Jianyin, 2013).

Table 3.1 summarizes the changing production costs and sales prices of bamboo laminated floorings in Anji from 1992 to 2012.

Case study 2: the evolution of Anji's bamboo product structure

Due to market changes, rising raw material and labour costs, and the emergence of new technologies and products, Anji's bamboo product structure has been changing continuously. The evolution of the product structure can be divided into the following five stages:

- 1980–1985: before industrial processing;
- 1986–1992: primary stage of industrial processing;
- 1993–1998: rapid development period of industrial processing;
- 1999–2005: expansion period; and
- 2006–now: transformation to a period of high-quality and environmentally friendly products.

In the detailed lists that follow for each period, those products whose names are shown in *italics* are new products for that period, and the sequence of the products goes from the most important products to the least important according to the mainstream market of the period.

1980–1985: BEFORE INDUSTRIAL PROCESSING. Raw bamboo (bamboo culms) → fresh bamboo shoots → bamboo scaffolding and daily-use products (farm tools, bamboo mats, handicrafts, brooms, baskets) → traditional papermaking→ traditional furniture (pole furniture) → medicine (Fig. 3.40).

1986–1992: PRIMARY STAGE OF INDUSTRIAL PROCESSING. *Bamboo mats* → *processed shoots* → fresh bamboo shoots → *bamboo curtains, chopsticks* → *sticks* → *flooring* → mats/curtains → *bamboo ply* → *processed bamboo poles* → scaffolding, bamboo products in daily use/handicrafts → *bamboo pulp* → traditional furniture → medicine (New products in italics) (Fig. 3.41).

1993–1998: RAPID DEVELOPMENT PERIOD OF INDUSTRIAL PROCESSING. Bamboo curtains → *bamboo carpets* →mats →bamboo floorings → processed shoots → fresh bamboo shoots → mats/curtains/ bamboo ply → processed poles → products in daily use → chopsticks → *laminated furniture* → *bamboo wood composite* → bamboo handicrafts → *charcoal/coal tar* → bamboo pulp → brooms → *bamboo leaf extraction (flavone products)* → *particle boards* → scaffolding → medicine (Fig. 3.42).

1999–2005: EXPANSION PERIOD. Bamboo floor → bamboo curtains → bamboo carpets → *bamboo mats (dining table mats, car seat mats and pillow mats)* → charcoal, coal tar and bamboo charcoal products (health products, medicines, pesticides) → *Pressed bamboo (other names describing the*

Table 3.1. Production costs and sales prices of bamboo laminated floorings (average of perpendicular and side laminated plates) in Anji, 1992–2012 (data from Zhu Zhaohua and Chen Jianyin, 2013).

Year	Raw material cost/m² (CNY)	Product cost/m² (CNY)	FOBᵃ price (CNY)	Exchange rate with US$
1992–1993	52.1	86.04	250	1:5.5
1995–1998	56.1	92.7	220	1:8.7
2001–2003	59.6	98.43	145	1:8.2
2006–2007	62.5	103.2	110	1:7.5
2010–2012	63	123	110.5	1:6.7

ᵃFree On Board.

Fig. 3.40. (a) Emerging bamboo shoots; (b) fresh shoots; (c) traditional baskets for shoulder carrying; (d) pots made from culm cross sections; (e) scaffolding; (f) medicines made from bamboo and bamboo juice; (g) bamboo rafts; and (h) traditional bamboo furniture.

Fig. 3.41. Bamboo products: (a) curtains; (b) processed bamboo shoots as leisure food; (c) chopsticks; (d) flooring panels; (e) panels for doors, walls and flooring; and (f) bamboo ply (Zhu Zhaohua).

same material include: bamboo-based fiber composite, bamboo scrimber, strand woven bamboo material, reconstituted bamboo material, etc., herein after all called pressed bamboo) panels →particle boards → processed shoots → fresh shoots → laminated furniture → mats/curtains/carpets → bamboo ply → *bamboo fibre products* → bamboo products in daily use → processed bamboo poles → medicines → bamboo sticks (toothpicks, chopsticks) → bamboo leaves → *bamboo veneer* → handicrafts → scaffolding → charcoal and charcoal products (such as health products, medicines, pesticides) (Fig.3.43).

2006–NOW: TRANSFORMATION TO PERIOD OF HIGH-QUALITY AND ENVIRONMENTALLY FRIENDLY PRODUCTS.   Pressed bamboo materials → *decorative boards* → *outdoor decking* → particle boards → bamboo curtains/carpets →laminated furniture → bamboo veneer → charcoal, coal tar and charcoal products → bamboo fibre products → bamboo mats (dining table mats, car seat mats and pillow mats) → bamboo composite panels → bamboo ply → processed shoots → fresh bamboo shoots → processed bamboo poles → bamboo leaf extraction products (flavone products) → bamboo leaves → *bamboo biofuel pellets* → bamboo products in

Fig. 3.42. Bamboo products: (a) mats; (b–c) laminated furniture; (d) wood composite block; (e) making charcoal bricks from bamboo sawdust; (f) charcoal; (g) bamboo leaf extracts; and (h) particle boards (Zhu Zhaohua).

Fig. 3.43. (a) Pressed bamboo; (b) bamboo fibre clothing; (c) bamboo veneer; (d) floorings made of pressed bamboo material (e) bamboo leaves being collected and removed; and (f) bamboo curtain/blinds (Zhu Zhaohua).

Fig. 3.44. (a) Bamboo biomass fuel pellets; (b) a building constructed of pressed bamboo; (c) outdoor decking made from bamboo; and (d) modern bamboo pole furniture (Zhu Zhaohua).

daily use (*bamboo tooth brushes*) → bamboo chopsticks → *modern bamboo pole furniture* → bamboo sticks → *pressed bamboo structures and buildings* → bamboo leaves → scaffolding → medicinal materials (Fig. 3.44).

3.3.4 Deep exploration of bamboo forest values – ecological and cultural products of bamboo forests

In the section above, we discussed bamboo products and their structure; these products came from the actual bamboo materials, such as stems, branches, leaves, rhizomes, roots, shoots, shoot sheaths, etc. The functions and roles of bamboo and bamboo forests can, however, be much more than a material resource. They possess rich ecological and cultural resources that could provide significant services to human beings and their lives, for example, beautiful landscapes and environment, fresh and clean air and water, and the conservation of watershed and slopes, as well as some ecological and cultural products based on the rich significance of bamboo in ecology and culture.

Bamboo ecological industry development

Many ecological products could be developed from bamboo forests and bring great ecological and economic benefits. Let us take the example of the bamboo ecological industry in Anji, Zhejiang Province. This is in its initial stage at the moment, but it has great potential.

A NEW GROWTH POINT FOR IMPROVING RURAL LIVELIHOOD AND INCREASING RURAL INCOME. Anji started to develop ecotourism at the end of the 1990s, and over 15 years of development, the sector has become a large-scale rural industry. Ecological tourism has become a new growth point for rural economy and farmers' income in

Fig. 3.45. Tortoise-shell bamboo (*Phyllostachys heterocycla*; syn. *Ph. edulis* var. *heterocycla*) (Zhu Zhaohua).

the county. In 2013, it attracted 104.403 million tourists and generated a total income of 10.231 billion CNY, the revenue of gate tickets for scenery spots was 160 million CNY, with a total number of over 40,000 people employed in the ecotourism and farm-stay sectors (Zhu Zhaohua and Chen Jianyin, 2013).

What attracted this many tourists to Anji? Besides the traditional cultures in this ancient county, the major attraction is the large areas of bamboo forests. There are 57,333 ha of Moso bamboo forest and 14,666 ha of forest with other species in Anji. Tourists like the bamboo land-scapes, clean air and water, tasty bamboo shoots, various other bamboo products and the long-established bamboo civilization in the area – the pure and simple folk life in the villages.

According to statistics, bamboo-related ecotourism generated a production value of 4.09 billion CNY in 2013, which accounted for 40.1% of the total tourism value. As for farm

stay, almost 100% of the tourists preferred the bamboo forest hostels. In 2013, over 2000 bam-boo forest households reached the requirements for receiving tourists and guests, and received certification for farm-stay management. They provide an accommodation capacity of more than 30,000 beds, which receive over 3 million people/year. The average annual net income of these households to date is more than 80,000 CNY (over US$13,300), and the top annual in-come exceeds 500,000 CNY.

The first village that developed farm stay is Daxi. In 2003, the village received 400,000 tourists. By 2013, the number of tourists to Daxi exceeded 1 million, and there were 166 house-holds in the village managing farm-stay busi-nesses, with the accommodation capacity reaching 3957 beds. The total production value generated through farm stays and tourism in 2013 reached 112.45 million CNY. Daxi has a total of 1560.5 ha of bamboo forests, 1000 ha

Fig. 3.46. A bamboo tourist reception and restaurant area in the Philippines (Zhu Zhaohua).

Fig. 3.47a,b. Bamboo structures for tourists in the tourism zone of Bolivia (Zhu Zhaohua).

of mixed forest of conifers and broadleaves and 67 ha of hickory [*Carya cathayensis*] forest. The total forest coverage is 85%.

Developing bamboo ecotourism has become a new way to generate livelihoods and incomes for the rural area. Yet the significance of ecotourism and farm stay in the area are far more than simply increasing rural income. First, they have increased employment opportunities in the rural area. Over 15,000 people in Anji were directly or indirectly engaged in farm-stay busi-

nesses in 2013, and these people were employed at home, while they could still manage their own farms and bamboo forests. Secondly, the development of the ecotourism sector has played an effective role in reducing the gaps between urban and rural areas. This process has promoted communication between the two types of area and therefore facilitated the elimination of the rural–urban divide, and enhanced social harmony. Thirdly, the development process of the ecotourism has brought investment and information from

Fig. 3.48. A floating restaurant constructed of bamboo in the Philippines (Zhu Zhaohua).

urban to the rural areas, and pushed forward the processes of rural modernization, civilization and improvement of the ecological environment. Fourthly, ecotourism has promoted the sales and increased the quality of agricultural and forestry products.

The above changes show that bamboo ecotourism and farm stay are deep explorations of the comprehensive values of the bamboo forests, and the value generated by the bamboo forests have far exceeded the value of the biomass product.

BAMBOO FOR THE ENVIRONMENT AND ECOLOGICAL SERVICES. Bamboo forests have strong soil and water control functions, and bamboo ecological construction projects may help to rehabilitate the local ecological environment. There are a certain number of middle, small and grass-sized bamboos, such as *Ph. aurea*, *Fargesia* spp., *Sasa argenteostriata*, *Indocalamus* spp., *Shibataea* spp., *S. chinensis* and *Chusquea* spp. that are suitable for this purpose. Amphipodial bamboo species, for example, *Pseudosasa amabilis* and *Guadua*

spp., are usually good at conserving soil owing to their special root systems. Large-sized species, such as Moso bamboo and *Dendrocalamus* spp. are also fine for water and soil conservation, and their effects could even be better when they are mixed with trees.

Bamboo forest can also play important roles in small-scale watershed conservation, river/lake bank protection; the rehabilitation of mined areas and degraded lands, landscape restoration, etc. In ecological projects, bamboo does not only bring fast and beneficial effects, but can also generate high-value products such as bamboo shoots and poles.

BAMBOO FOR LANDSCAPING AND ORNAMENTATION. Bamboo's elegant form and fast greening effects have been recognized by more and more people and so bamboos are becoming more and more important in gardening and landscaping. In many countries, especially South-east Asian countries, such as China, Japan, South Korea, Thailand and Vietnam, bamboos are indispensable

Fig. 3.49. *Indocalamus decorus* as ornamental landscaping (Zhu Zhaohua).

parts of urban greening and landscape config-uration. Bamboos are an important cultural and stylish element in the gardening arts of these countries, and they are used in landscap-ing, as fences, in ornamental gardens and vari-ous other types of structures, buildings and bridges. In the Philippines, bamboo is also used to build nice looking, local stylish ferry boats (Zhu Zhaohua, 2012c).

The landscaping and ornamental func-tions of bamboo could become a very important value and have great potential for development in the future. For example, an ornamental bam-boo garden located in Lin'an, Zhejiang, China, had an area of 3.3 ha of bamboo nursery, which produces over 1.5 million seedlings of *Arundi-naria argenteostriata* (a grass-sized bamboo), and 1.2 million seedlings of other species. In a 23 ha plantation, the garden produces 172,500 mother stocks of bamboo for ornamental pur-poses in urban areas. The annual income of the garden from growing bamboo seedlings and mother stocks was 9 to 10 million CNY/year. These seedlings and mother stocks may be used for urban greening or further propagation in other nurseries. If all of these seedlings and

mother stocks are all utilized for urban land-scaping and ornamentation, they will generate enormous value.

The development of bamboo cultural products

Bamboo has great cultural significance. It can be utilized for various cultural products with high effectiveness and high values. Bamboo's cultural values may not have caught enough attention at the moment, but once developed, the effects may be beyond expectations. Let us look into several 'small cases'.

From 1984 to 1986, The County Govern-ment of Anji collaborated with the Subtropical Forestry Research Institute of the Chinese Acad-emy of Forestry in building a bamboo botanical garden. The garden collected over 280 bamboo species. In the following years, the garden intro-duced ornamental stones carved with historic literature and poems about bamboos, and artis-tic bamboo bonsai and bamboo handicraft shops. In 2000, a Bamboo Museum was established in the garden to raise people's awareness of bam-boo culture. In 2006, two pairs of giant pandas

Fig. 3.50. A bamboo structure constructed in the bamboo forests of Anji, Zhejiang Province, as a site for shooting the Chinese movie – *Ye Yan* (*The Banquet*, 夜宴) (Zhu Zhaohua).

were introduced and became a new attraction. As a result, this garden did not only popularize bamboo knowledge and culture, but also provided large amounts of mother stocks for other gardens and ornamental projects, and generated considerable income. To date, the garden is an attraction to 500,000 tourists every year and the ticket income has accumulated to over 15 million CNY.

Taking advantage of the special beauty of its bamboo landscape, Anji has constructed two movie shooting sites in the bamboo forests. The well-known movies *Crouching Tiger, Hidden Dragon* (臥虎藏龍) and *The Banquet* (*Ye Yan*, 夜宴) both selected Anji's bamboo forests as their filming sites. The beautiful bamboo forest scenes in the movies attracted large numbers of tourists to the sites, and annual visits to the two sites are 1.5 to 2 million, thus generating great economic value for the county.

Here is another example of success. The Baishui Stream, a famous tourism attraction in Lin'an, Zhejiang Province, was well known for its steep valley, green mountains covered with tall and straight Moso bamboo and an appealing waterfall scenery spot. Besides the beautiful scenery and comfortable environment, tourists could also enjoy a special feast of bamboo shoots and a bamboo cultural night, with singing, dancing, playing bamboo instruments and

Fig. 3.51. Artificially modelled Moso bamboo (*Phyllostachys heterocycla* var. *pubescens*) for tourism purposes in Anji, Zhejiang Province, China (Zhu Zhaohua).

appreciating folk performances, blending with the local farmers and feeling soaked in the pureness and simplicities of the rural environment and nature. A local farmer expressed satisfaction with the values brought by the bamboo culture industry: 'It's unbelievable that our folk bamboo stories could attract so many tourists and generate this much income.' Every year around 200,000 people visit the site, and some of them stay in the farmhouses for days and even months.

> *The bamboo culture industry covers a large range of areas, but there are still great blank spaces waiting for people to explore.*

3.4   Demonstration Sites: the Guarantee for Success

In any country or region in the world, when initiating a scaled bamboo industry, a precondition

Fig. 3.52. The Baishui Stream Scenery Site, a rural ecotourism site in Lin'an, Zhejiang Province, China (Zhu Zhaohua).

is to accumulate knowledge and learn lessons from the practical experiences of others. A more important initial stage is to trial on a small scale, and when this demonstration/pilot project achieves success, the model can be introduced and extended to other areas for large-scale development. The small-scale pilot site needs to be provided with the necessary technologies, policies and funds, and have the participation of local stakeholders, to guarantee success. The provision of these conditions is what we understand as the construction process of the demonstration site. The construction of a demonstration site is to avoid blind investment and detours and, what is more, accumulate practical experience and build confidence among the stakeholders, so as to appeal for more investors, policy makers and communities to engage in the industry.

Fig. 3.53. A folk performance by local farmers at the Baishui Stream Scenery Site in Lin'an, Zhejiang Province, China (Zhu Zhaohua).

3.4.1 Conditions for a demonstration site

The purposes of a demonstration site depend on the actual needs of the local bamboo industry development. For example, if the local bamboo resources are insufficient for industrial processing, the first demonstration site to be considered should be a bamboo resource cultivation base, but if the resources are sufficient, a bamboo processing demonstration site could be constructed. At the same time, a demonstration for high-yielding and high-quality bamboo forest management should be in place to guarantee the sustainability of the processing industry. Below, we list the key conditions to constructing a demonstration site, and in the rest of Section 3.4, we discuss some of these conditions and approaches further and provide a case study.

- Selection of a suitable area: select a concentrated area of the bamboo resources: to provide a convenient and sufficient raw material supply to the bamboo processing industry, which at the same time could allow more and various processing companies to join in the future.
- Presence of a convenient transportation system: this is to allow access for interested

people to visit and learn, so as to enlarge the impacts of the demonstration site while, of course, providing convenient transportation for raw material supply and market access.

- Local government attention and support: these are indispensable for the success of a demonstration site. A preferential condition for the sector to grow would be that the major officials in the local government are quite familiar with and interested in the bamboo sector, or are active advocators and supporters of bamboo development at the local level, so that they could provide support in policy and other aspects when there is a need. A favourable policy environment could be key for bamboo enterprises to grow.
- Wide representativeness: the bamboo species, the standing condition of the forests, the local rural socio-economic conditions, etc. should be representative at the local, or national or regional level. This will allow a wider extension space for the demonstration site.
- Active participation of local communities: after awareness raising and training activities, the communities in the demonstration area should be willing to engage in or support the construction of the demonstration

site, and at least some of the community members should understand the significance of the demonstration site well.

3.4.2 Multi-participation – full functioning of the demonstration site

We have mentioned above the importance of government participation, yet the most important factor for a demonstration site should be the participation of the enterprises concerned and the local communities. The combination of the two stakeholders will create a win–win situation, which is essential for the success of the demonstration. The participation of scientists and technicians is also important, as they play a major role in providing training to the local communities, allowing information exchange and normalizing and standardizing production and products to guarantee the overall quality control of the demonstration site. In a word, the government, enterprises, scientists and technicians, and the local communities (bamboo famers) and other relevant sectors and organizations should work together closely to contribute to the construction and solidarity of the demonstration site.

3.4.3 Different demonstration sites pilot different stages of development and guarantee the sustainability of the bamboo sector

People usually pay a lot of attention to the construction of a demonstration site for the bamboo sector in the initial stages, but when the bamboo industry successfully makes its first step, people may ignore the need for a higher level demonstration site to guide the whole industry to achieve further development. This may result in the loss of motivation and competitiveness for new developments, and finally stop the growth of the whole bamboo sector.

An industry may face different challenges and conflicts at different stages of development, and the developers should be able to see the prospect for future development according to the trends and changing situation, predict new challenges and establish new demonstration sites to trial and pilot future actions, so that the new challenges are overcome and the bamboo industry brought to a new level. Thus, demonstration sites should be established according to the needs of different development stages and they may cover a wide range of issues, such as: the rehabilitation of low-yielding and low-valued natural bamboo forests; the cultivation and management of high-yielding bamboo plantations; the operation and management of the processing industries; research and development on new technologies and products; the establishment and management of professional markets

Fig. 3.54a,b. Giant pandas eating bamboos and bamboo shoots in Sichuan Province, China (Jin Wei and Zhu Zhaohua).

for raw material supply and products; training and capability building of local communities, etc. In short, different demonstration sites should be constructed at different stages of development to deal with the new conflicts and challenges; the policies and the major acting stakeholder entities should also be changed according to need.

3.4.4 Case study: bamboo shoot industry development in Lin'an, Zhejiang Province – a case of demonstration site construction in the different development stages

> **A fine example has boundless power!**

We will take the example of bamboo shoot industry development in Lin'an (Lin'an Forestry Bureau, 2014; Wang Anguo, 2014) to show the importance of demonstration sites. For information on the bamboo resources in Lin'an, refer to Section 3.2.4.

Feasibility study

In Lin'an, there are many bamboo species that are fine for shoot (food) purposes, such as *Ph. praecox*, *Ph. vivax*, *Ph. nuda*, Moso bamboo, etc. Most rural households traditionally had a small area of bamboo stands in front of or behind their houses and most of the harvested shoots were consumed at home, with a small surplus amount sold at local markets. In order to meet the national

goal of quadrupling the national economy by 2000, in 1982 and 1983, the local forestry bureau of Lin'an collaborated with researchers and experts and carried out an investigation that looked for products and economic crops that had potential in increasing the farmers' income over a short period. Through the investigation, they found that some households cultivating *Ph. praecox* and *Ph. vivax* and selling their shoots were able to increase their incomes in a short period of time. In 1983, the experts proposed to the local government a plan to develop the bamboo shoot sector; the plan suggested expanding the existing 2000 ha of *Ph. praecox* to 6700 ha in 10 years, and also making the bamboo shoot sector a key sector for achieving the goal of quadrupling the local economy. The plan received great attention from the major leaders of the local government and in 1985, policies were made to initiate and provide strong support to the planting and expansion of bamboo plantation for shoot purposes.

Development results from 1986 to 2013

By the year 1991, the forest area of *Ph. praecox* and *Ph. vivax* had expanded to 7059 ha, which was much more than the 10 year target of 6700 ha. Up to 2005, the total area of bamboo forests in Lin'an (for shoot purposes) was 22,000 ha (see Fig. 3.55). This figure remained almost the same in the following years, with not much distinguishable expansion happening up to 2013. Fresh shoot production increased from 1420 tons in 1983 to 197,000 tons in 2005

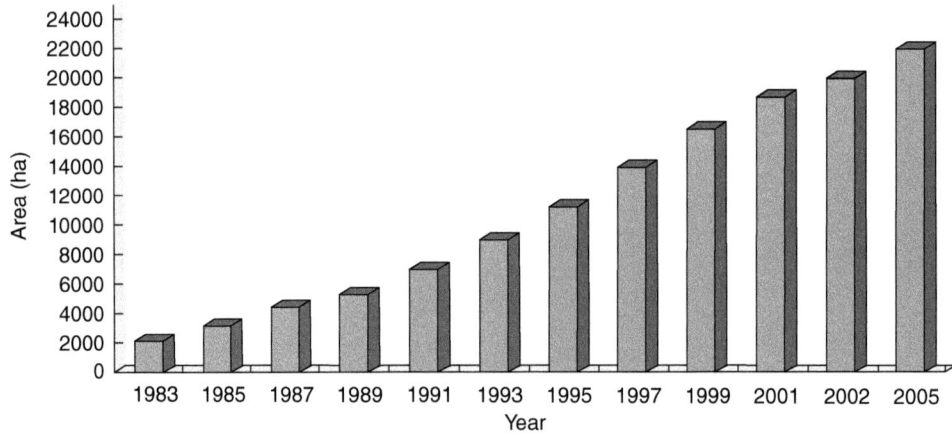

Fig. 3.55. Growth of bamboo shoot plantation area in Lin'an, Zheijiang Province, China, from 1983 to 2005 (Zhu Zhaohua, 2007b).

(see Fig. 3.56) and to 250,000 tons in 2013, and the value increased from 219 thousand CNY in 1983 to 560 million CNY in 2005 (see Fig. 3.57) and to 900 million CNY in 2013. The value of bamboo shoot processing increased from 97.15 million CNY in 1997 to 620 million CNY in 2005 (see Fig. 3.58) and to 3.184 billion CNY in 2013. The total production value of the bamboo shoot industry in Lin'an increased from 400 million CNY in 1997 to 1.17 billion CNY in 2005 and 4.084 billion CNY in 2013, and the export volume of bamboo shoots increased from 1287 tons in 1996 to 41,500 tons in 2005 (see Fig. 3.59) (Zhu Zhaohua, 2007b; Zhu Zhaohua and Wang Anguo, 2014).

Some 60% of the rural households in Lin'an were engaged in bamboo shoot production, which was estimated to be more than 70,000 households. The average annual net income per capita in the rural areas increased from 552 CNY (US$339) in 1985 to 7260 CNY (US$897) in 2005 and 17,561 CNY (US$2832) in 2013 (Zhu Zhaohua and Wang Anguo, 2014). Taking Gaohong Town in Lin'an as an example, the percentage of bamboo shoot income compared with the total per capita income was respectively 19.1% in

1990, 53% in 1995, 70% in 1996, 55.2% in 1998, 45.3% in 2002 and 20–30% in 2013. The reason for the reduction in percentage per capita income since 1996 is that when the famers' incomes were sufficient, they started to invest in other profitable industries. The major industry in Gaohong Town is the production of energy-saving electric lamps. Some farmers became owners of the lamp factories, some became shareholders and the others became workers in the factories. So although the annual income from the bamboo shoot industry has been increasing, the percentage of the total annual income from shoot production has been decreasing owing to the additional income from other businesses the farmers are managing. Hence, the bamboo shoot industry has not only alleviated poverty in Lin'an's rural areas, it has also helped the farmers to accumulate basic wealth and capital for the development of other rural industries, and laid the basis for a more reasonable, secured and sustainable rural economy and industry structure. Table 3.2 summarizes the statistics of bamboo shoot production volume and value in relation to rural per capita income for Gaohong Town from 1990 to 2002.

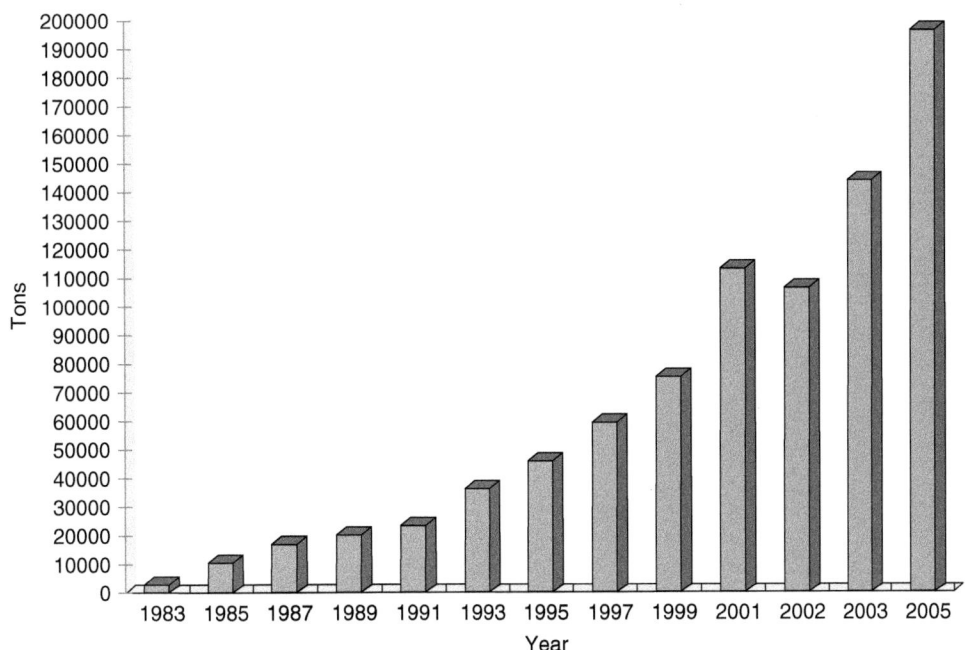

Fig. 3.56. Growth of bamboo shoot production in Lin'an, Zheijiang Province, China, from 1983 to 2005 (Zhu Zhaohua, 2007b).

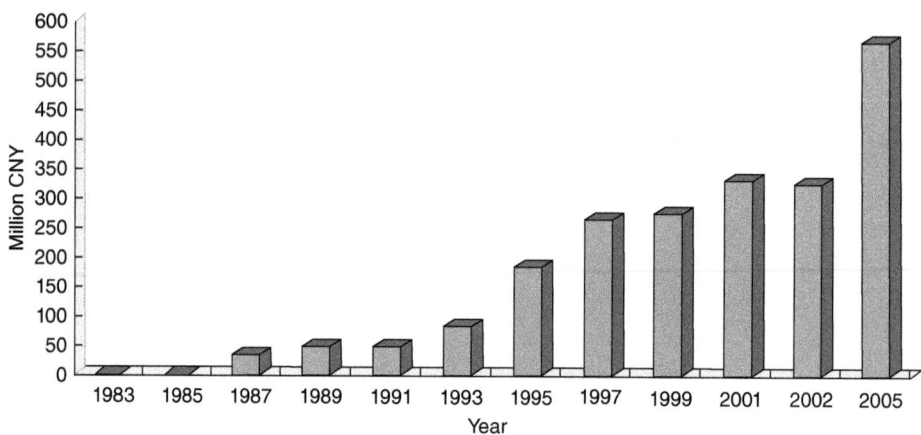

Fig. 3.57. Growth of the value of fresh bamboo shoot production in Lin'an, Zheijiang Province, China, from 1983 to 2005 (Zhu Zhaohua, 2007b).

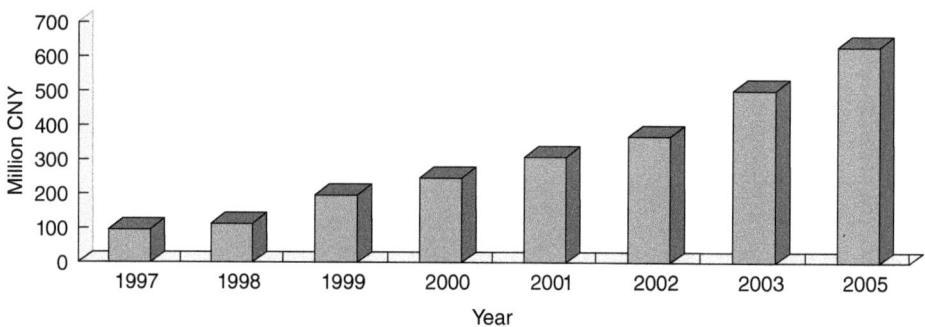

Fig. 3.58. Growth of the value of bamboo shoot processing in Lin'an, Zheijiang Province, China, from 1997 to 2005 (Zhu Zhaohua, 2007b).

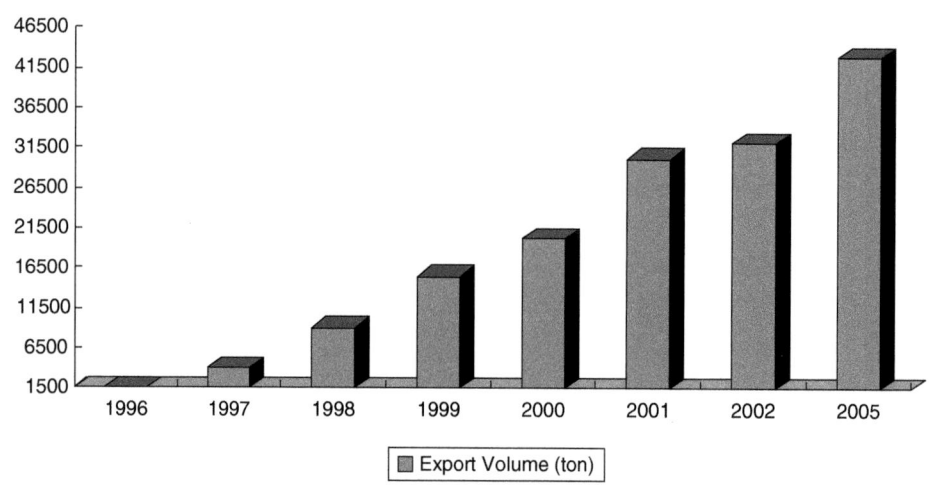

Fig. 3.59. Growth of the bamboo shoot export volume in Lin'an, Zheijiang Province, China, from 1996 to 2005 (Zhu Zhaohua, 2007b).

Table 3.2. Statistics of bamboo shoot production volume and value for Gaohong Town, Lin'an, Zheijiang Province, China, 1990–2002 (from Zhu Zhaohua, 2007b).

Year	Production value (million CNY)	Production volume (tons)	Rural per capita net income (CNY)	Bamboo shoot income (%)
1990	1.80	690	931	19.1
1991	2.30	810	–	–
1992	3.50	1050	–	–
1993	5.80	1260	–	–
1994	6.80	1960	2111	32
1995	17.00	3024	3173	53.1
1996	26.18	4133	3641	70
1997	21.32	5838	3792	67.8
1998	25.35	6002	4580	55.2
1999	20.18	6085	4270	47
2000	21.68	6853	4510	47.7
2001	25.80	6118	–	–
2002	23.96	6670	5183	45.3

What are the reasons for such excellent progress? Demonstration sites played an important role

The success of Lin'an bamboo industry should be attributed to:

- The multi-participation of government, enterprises, scientists and technicians, and local communities; they have all been dedicated and responsible for their roles.
- The leading roles of the demonstration sites, which have successfully balanced the development of the primary, secondary and tertiary industries, and achieved a multi-win situation.

The following sections give analyses of the different development stages (periods) of Lin'an's bamboo shoot industry.

1985–1991: RESOURCE DEVELOPMENT THROUGH DEMONSTRATION BY PROFESSIONAL BAMBOO HOUSEHOLDS. The most important mission in this period was to raise the awareness of the farmers of the importance of the bamboo shoot industry in improving their livelihoods. Technical training sessions were provided to the households that recognized the importance of bamboo shoots the earliest and had actively participated in the planting of bamboo. These households were allowed to utilize the collective-owned degraded lands in the village. Each hectare of new bamboo plantation

could receive 300 CNY/year from the government as subsidies. From 1986 to 1988, the local government input 1 million CNY/year for the expansion of bamboo plantations for shoot purposes. Households that grew the best quality bamboos and shoots were recognized as 'Demonstration Households', and were prioritized for technical training and services. The demonstration households were not only honoured by their title and certificates, but also rewarded with cash and fertilizers. By 1992, the number of bamboo shoot demonstration households had exceeded 300 and a group of farmers even acquired the professional title of 'rural technician'. The number of demonstration villages reached 102 and have had a great impact on the whole rural area of Lin'an.

In 1989, the policy for subsidizing bamboo plantations for shoot purposes was stopped, because the first group of bamboo farmers who pioneered bamboo shoot forest management in 1984 and 1985 had obviously benefited through the bamboo shoot production. By then, most local farmer households believed that bamboo management could generate a considerable income, so there was no need for further incentives from local government. The farmers began to invest in bamboo plantations with their own resources and the bamboo forests expanded fast. In Gaohong Town, where most farmer households managed bamboo forests for shoot production, the total income from bamboo shoots increased from 2.4 million CNY in 1997 to 27 million CNY

in 2002. A model village named Chenjiaba expanded its bamboo plantations from 7 ha in
1990 to 22.2 ha in 1995. In 1996, the total income brought by bamboo shoot production to
this village was 1.028 million CNY, and with a
population of 102, the per capita income was
over 10,000 CNY. By 2002, the bamboo plantations of the village had expanded to 80 ha, and
the village had become rich in the local area. In
2003, there were 11,000 households in Lin'an
whose bamboo shoot income exceeded 10,000
CNY, over 4000 where it exceeded 20,000 CNY,
over 180 where it exceeded 50,000 CNY, ten
where it exceeded 100,000 CNY and three where
it exceeded 200,000 CNY.

**1992–2003: PROMOTE THE BAMBOO SHOOT INDUSTRY
THROUGH SCALING UP HIGH-YIELDING AND HIGH-QUALITY
DEMONSTRATION FORESTS AND DEMONSTRATION
PROCESSING FACTORIES.** Due to the fast expansion of the bamboo plantation areas and the increase of bamboo shoot production, the bamboo
shoot sector development in Lin'an faced new
challenges:

- The bamboo plantations had been expanding too fast, resulting in large areas of
  monoculture, which poses high risks.
- Despite the fast expansion, the management of the bamboo plantations had not
  been intensified, resulting in a number of
  low-yielding bamboo plantations.
- Because of the sudden increase of supply,
  the market price of the fresh bamboo shoots
  had reduced, and large amounts of fresh
  shoots were in urgent need of immediate
  processing.

To enable it to deal with the above challenges,
the local government took a number of measures to, on one hand, slow down bamboo
plantation expansion and, on the other hand,
establish high-yielding demonstration sites
and demonstration shoot processing factories
to guide the sector on to a route of healthier
development.

In order to increase unit production and the
shoot quality, non-polluted and green products
needed to be produced. This required demonstration over larger areas and higher standards. In
2000, the local government decided to organize
the bamboo farmers in two villages in Taihuyuan
Town – Yangling and Qingshan – to set up two
large-scale, high-yielding and high-standard
bamboo shoot forest demonstration sites of, respectively, 670 and 80 ha. The Vice Mayor of Lin'an,
who was responsible for agriculture and rural
development, and the Party Secretary of Lin'an
were both assigned to be responsible for the two
demonstration sites. An expert group was established to provide standardized management
technologies for the sites (including fertilizing,
watering, density control, 'coverage' (mulching)
technologies to encourage early shooting, pest
and disease control, etc.), standardized bamboo
shoot quality (including pesticide residues, heavy
metal content, etc.), and standardized unit production, etc. These two demonstration sites had a
great impact on and improved the level of bamboo shoot forest management in Lin'an.

In 1984, Lin'an started to introduce a new
technology – 'coverage' technology – for bamboo
shoot forests to promote shooting. It was first applied by the demonstration households in their
own demonstration plantations, where farmers
can use the technology to control shooting time,
so as to provide off-season shoots for the market.
The technology has advanced the shooting time
from middle March to November in the year before, and has increased the shooting period from
1.5 months to 5–6 months. It has improved not
only the productivity but also the value of the
shoots. Normally, fresh shoots were available only
in the natural shooting season (March–April),
when the market price was 2–3 CNY/kg; in contrast, off-season shoots, which were put on to the
market from November to February, had a market price of 10–20 CNY/kg. In his demonstration
household in Xia village, Zhu Yourong used the
coverage technology to manage 0.1 ha of bamboo forest, and his income from shoot production
in 1998 was 82,000 CNY (then US$100,000/ha),
which was the highest income record in Lin'an's
bamboo shoot production history.

By 2003, the coverage technology had been
extended to 3400 ha of bamboo forest. Statistics
showed that from 1991 to 2003, this technology had increased Lin'an farmers' income by
1.482 billion CNY. In 1993 and 1994, with the
higher motivation of the farmers in an expanding bamboo plantation area, the bamboo shoot
production of Lin'an reached an even higher
level, and the market price of fresh shoots began
to fall, especially in the natural shooting period
from March to April.

Fig. 3.60. Shoots of *Phyllostachys praecox* emerging from the soil after applying 'coverage' (mulching) technology (Wang Anguo).

The government's supporting priority next shifted from the extension of the bamboo plantation area to bamboo shoot processing, and enterprises were encouraged to invest in shoot processing. In 1994, with investments introduced from Taiwan, Lin'an established its first joint venture for shoot processing. The demonstration and pioneering role of this company made it one of the earliest leading companies in promoting the local shoot processing industry. From 1995 to 1997, the Lin'an government provided a soft loan of 1.5 million CNY/year (the government pays the interest) to support the development of the bamboo shoot processing industry. Together with the successful demonstration of shoot processing enterprises developed in the earlier period (1994–1995), and the local government's supporting policies, Lin'an's bamboo processing enterprises increased to 36 in 1997, and their processing capacity reached 22,000 tons. Up to 2008, the number of factories increased to 48 and the processing capacity to 60,000 tons. Furthermore, the processing factories of Lin'an not

only consumed local shoots, but also shoots from other locations.

2003–2008: CONSTRUCT FRESH SHOOT MARKETS AND DEVELOP DOMESTIC AND OVERSEAS MARKETS. With the increase of fresh bamboo shoot production, the traditional method of bamboo shoot selling – middleman purchasing and transportation – seemed rather inefficient. In 2002, the experts proposed that specialized fresh shoot markets be established to promote sales. A demonstration fresh shoot market was established in Gaohong Town, the land was provided by the local government and an enterprise was introduced for management. While local farmers brought their fresh shoots to the market right after harvesting, the middlemen could directly purchase from the market and transport them to medium and large cities, such as Hangzhou, Shanghai and Nanjing, etc. People in these cities could buy fresh shoots harvested early in the morning of the day. This specialized market has largely promoted the efficiency of fresh shoot trade and reduced costs. Farmers could sell their shoots to

Fig. 3.61. Harvesting winter shoots in the snow in Lin'an, Zhejiang Province, China (Wang Anguo).

middleman who offered highest price and get more income.

The success resulting from the demonstration fresh shoot market in Gaohong Town encouraged the other ten bamboo-shoot producing towns in Lin'an to establish specialized fresh shoot markets. Up to 2003 and 2004, there were in total 15 fresh bamboo shoot markets in the whole county, of which three were located in Taihuyuan Town. The daily sales of one market could reach 100 to 150 tons during the period from middle November to middle April, and the highest daily market sales reached 300 tons. With the facilities offered by these markets, some farmers were able to participate in the sales and transportation of fresh shoots. According to statistics, to date, over 5000 farmers in Lin'an are engaged in the sale of fresh shoots.

Besides the construction of markets within Lin'an, the local government also established

Fig. 3.62. A demonstration bamboo plantation for shoot production purposes in Lin'an, Zhejiang Province, China (Zhu Zhaohua).

pilot markets in Shanghai and Suzhou to sell Lin'an-branded shoots. These markets were run by the bamboo farmers themselves and successfully promoted the Lin'an brand.

As a result of the impacts of the demonstration markets in Shanghai and Suzhou, further steps were taken to develop markets in nine bamboo shoot producing regions in China and other countries. In order to develop international markets, the Lin'an Bamboo Shoot Industry Society organized major processing enterprises to make a study tour to Japan, which was the largest foreign market for bamboo shoots. To expand markets to the vast area north of the Yangtze River, a delegation led by the Zhejiang Provincial Department of Forestry, with government representatives and participants from enterprises and major shoot sales households, went to major cities such as Beijing, Tianjin, Xi'an and Harbin to promote bamboo shoot markets; they also gave information on the nutritional value of bamboo shoots and introduced shoot cooking methods to these cities. These promotional activities soon opened up the markets in north China, and continuous large demands from these markets kept the prices of bamboo shoot products stable and increasing steadily.

2008–NOW: DEVELOPMENT OF ORGANIC BAMBOO SHOOT PRODUCTS AND SUSTAINABLE MANAGEMENT OF THE BAMBOO FORESTS. Any industry, during the course of its development, may face various challenges and difficulties. After 20 years of development, the new challenges that the Lin'an's bamboo shoot industry is facing now are:

- soil acidification and productivity reduction after years of intensive management of the forest;
- increased vulnerability to diseases and pests due to large areas of monoculture; and
- higher requirements for quality and standards due to increased customer concerns on food security.

Fig. 3.63a,b,c. Photos from a bamboo shoot market in Lin'an, Zhejiang Province, China (Zhu Zhaohua).

Demonstration is still an indispensable way to deal with these challenges. For tackling the degradation of the soil, Lin'an Forestry Bureau cooperated with fertilizer supplying enterprises and demonstration households in trialling soil tests and prescribe suitable organic fertilizer in selected demonstration forests. Through tests and trials, the prescription organic compound fertilizer achieved great results. Both the productivity and the quality of bamboo shoots showed distinct improvements within 2 years.

Biological pest and disease control methods were also introduced to reduce or replace pesticides. Strict standards were applied for bamboo shoot products in order to access quality certifications and build local high-quality brands. By 2008, three bamboo shoot processing enterprises had accessed China's national QS (quality and safety) certification. Many products from Lin'an have also accessed international standards, such as ISO 9001 (the International Organization for Standardization's quality management system (QMS) standard for organizations), HACCP (a hazard analysis and critical control points certification system for keeping products safe during production), GAP (good agricultural practice certification) and JONA (certification by the Japan Organic and Natural Foods Association).

The experiences of Lin'an show us the importance of demonstration sites. Trials at these sites may provide experiences for future wider and large-scale development, may prevent failure and detours, and may strengthen the motivation of stakeholders to engage in and support the bamboo sector and build their confidence. The construction of demonstration sites is thus important for successful industrial development.

3.5 Policy Support: Key to Sustainable Development of the Bamboo Sector

The right policies are, in most cases, more important than pure financial support.

Fig. 3.64. Boiled Moso bamboo (*Phyllostachys heterocycla* var. *pubescens*) shoots ready for slicing in Lin'an, Zhejiang Province, China (Zhu Zhaohua).

There are many factors affecting the decision making of an investor looking to invest in the bamboo sector, but two major factors are: (i) the local bamboo resources, which have been discussed in previous sections; and (ii) the investment environment. The most important condition of the second factor, the investment environment, would be whether or not the local policies and/or regulations are positive.

The bamboo industry is affected by government policies in many aspects, such as resource management, land ownership, taxation, investment, the workforce, etc. Furthermore, policy is one of the key factors affecting the success of the bamboo industry. To ensure sustainable development of the bamboo industry, the local government and policy makers should provide systematic and consistent policies according to the particular features of the bamboo industry and adopt different strategies at different development stages.

3.5.1 Consistency of country policies

Many countries have already identified a series of policies for forest management and industry

Fig. 3.65. The processing factory of Hanghzou Kangxin Food Co., Ltd. in Lin'an, Zhejiang Province, China (Zhu Zhaohua).

Fig. 3.66. Pickled bamboo shoots in packages in Lin'an, Zhejiang Province, China (Jin Wei).

development. Here we need to discuss whether the policies are contradictory to each other.

For example, in some Eastern African countries, such as Tanzania and Kenya, most bamboo resources are distributed in natural forests, and according to national policy, the natural forest is protected, therefore logging is severely forbidden, including that of bamboo (Zhu Zhaohua, 2001). The largest difference between the bamboo forest and the other forests is that it has a strong renewal capacity, and that reasonable harvesting will increase this renewal capacity and production. On the contrary, no harvesting may result in the degradation and reduction of productivity of the bamboo forest.

In some Latin American countries, such as Colombia, the natural bamboo forests are mainly composed of the beautiful *G. angustifolia*. Because bamboo plants like water, most of these forests are distributed alongside the rivers and small watersheds. According to the statistics of the Colombian Ministry of Environment, natural forests by riversides and watersheds are protected to conserve soil and water, and bamboo forests are included in this protection (Zhu Zhaohua, 2005d). Although the other policies of these countries support bamboo-based development, if they do not make adjustments to their

natural bamboo forest management policies, the local bamboo industry may be hindered by lack of resources, and while bamboo could be planted, plantations take years to mature (Zhu Zhaohua, 2005a).

Another example is that in many countries of Asia, Africa and Latin America, bamboos grow naturally in the forest, but they are usually mixed with other trees. These countries have clear property policies about timber trees, but no policy for non-timber forest resources such as bamboos and rattans. In Huaphanh, Laos, most forests were owned by local communities, and the bamboo culms and shoots in the forests can be freely harvested by anyone, with no need for permission. The inappropriate cutting and harvesting that was done resulted in serious degradation of the bamboo forests.

There used to be the similar cases in China. For instance, Baisha village in Lin'an, Zhejiang Province used to be a typical mountainous poor village. The village forest coverage was 93%, and a small-sized bamboo specie for shoot purposes – *Ph. nuda*, grew naturally in the forests. Before 1984, when there was no identified property ownership for the bamboo stands, every May, the local farmers harvested as many bamboo shoots as possible, and as a result of overharvesting, the productivity of the

Fig. 3.67. Guadua bamboo growing on the banks of a river in Colombia (Zhu Zhaohua).

Fig. 3.68. Guadua forests distributed alongside small watersheds in Colombia (Zhu Zhaohua).

bamboo forests was reduced. However, since 1984, the natural forests have been contracted to households, including the stands of *Ph. nuda*. From 1991, technicians were sent by the local Forestry Bureau to provide training to farmers on management technologies for bamboo forests, including their proper density and harvesting. After this training, the natural bamboo forests soon recovered their productivity, and the per capita income of the village from bamboo shoots increased from 150.3 CNY in 1991 to 681 CNY in 1995, which was 4.53 times that in 1991 (Zhu Zhaohua, 2007a).

The above three examples, the first two concerning resource management and the third property ownership, are very important policies for bamboo sector development, though there are many other policies on other matters that also affect the healthy development of the bamboo industry. Missing, deficient or contradictory policies may hamper the industry's healthy development. Such problems happen when the related government sectors and policy makers do not share a common knowledge of bamboo industry development, or lack coordination.

China has many successful experiences in this respect. In China, when a local government is identifying a bamboo industry development plan, it will not only recognize a series of figures as objectives, but also measures to ensure that the objectives are achieved. Besides policies at national and provincial levels, the local government will identify improvement measures or new regulations according to local conditions, to guarantee the success of the plan. These new measures and regulations are implemented under the direct leadership of the mayor or secretary of the party committee, with coordinated actions by the different but related government departments. In this way, the policies can be guaranteed to be systematic and consistent, and their implementation unified and coordinated.

3.5.2   Policy making according to the natural rules and features of the bamboo industry

Many people think that the bamboo industry is easy to develop because it does not require high

technology or huge investment. This is a misunderstanding.

- First, the bamboo industry covers many fields – it is comprehensive and involves forestry, agriculture, biodiversity conservation and development, ecosystem construction, industrial processing and marketing.

- Second, the bamboo industry involves various groups of people in society, including farmers, entrepreneurs, workers and businessmen in domestic and foreign markets. Policy making should consider the benefits of and provide inspiration for the participation of all stakeholders in order to create a win–win situation in which the stakeholders are interdependent.

- Third, the bamboo industry has a comparatively longer 'cycle period', as even though bamboo grows faster than trees and can be harvested more quickly, the same area of bamboo forest can produce more over a period of time than that of trees. In comparison with other agricultural crops, fisheries or animal husbandry, the cycle period is also longer. If we calculate the cycle period from planting bamboo to products on the market, the cycle could be even longer. Therefore, the stabilization of the government policies could be an important condition for the sector to grow.

In Anji and Lin'an, where the bamboo industry is well developed, local government policies have been kept consistent over the long term of industrial development; whenever a new government is in place, there is rarely a change to the policies, although improvements are made on the basis of the previous government policies. Of course, this does not mean that all policies should remain the same in the long term. When a policy is no longer appropriate, it should be adjusted or removed.

An example is the bamboo agriculture fund of the Anji government. In the late 1980s, the local government began to charge a bamboo agriculture fund fee, which was 6% of the profits of the processing enterprises. The fund was used for providing services to the local farmers and enterprises, such as technical training, the construction of demonstration sites and forest roads, development of new products, etc. At the time, government revenue was quite limited, and this measure played an important role in driving the development of the bamboo industry. However, with fast economic growth in Anji, in 2007, the government's revenue was sufficient to support the bamboo industry, so the fee was alleviated. From 2000, the 8% tax on agricultural speciality products was also cancelled, so that by 2007, all fees and taxes on bamboo farmers had been removed.

Preferential policies corresponding to the needs of development at different stages developed. Case study: Anji bamboo industry development

The policy 'shoes' must meet the 'feet' of the growing bamboo sector.

The bamboo sector in a country or a region may go through a development course from zero to small scale, from small scale to large scale, from primary products to high-end products and from low value adding to high value adding products. In the different development stages, the policy making priorities should be different. Here, we will take again Anji's policy evolution as an example. Much of this section is based on an account of the development of the Anji bamboo sector compiled by Anji Forestry Bureau (2002), a research project evaluation of the bamboo industry's impact on rural development in Anji and the 2014 Yearbook of Anji Statistics from Anji Forestry Bureau (2015).

Since China's reform and opening up policy was implemented (from 1978 to now), Anji's bamboo industry development has gone through 36 years. The county has a long established history of bamboo cultivation. The earliest record of bamboo cultivation in Anji can be found in the Tang dynasty (AD 618 to 907); by then, Anji was already known as a 'Bamboo Hometown'. By 1978, the bamboo forest area in Anji was 54,822 ha, which was 29.07% of the county's territory area and 60% of the total forest area. Local communities had rich experiences in bamboo cultivation and utilization, so, Anji's bamboo sector did not start from zero. After the reform and opening up policy of 1978, Anji's bamboo sector experienced revolutionary changes: bamboo processing was transferred from traditional

manual work to mechanical industrialized processing; bamboo forest management was also transferred from traditional private management to large-scale industrial management, and further to scientific, sustainable modes. These changes could not take place without the guidance and supports of a series of policies.

The changes in Anji's bamboo sector over the past 35 years are summarized in Table 3.3, and over this period the sector can be divided into the four stages (I–IV) which are described in the four subsections below.

STAGE I. BAMBOO RESOURCE DEVELOPMENT — IMPROVEMENT IN STANDING QUALITY AND PRODUCTIVITY — AND TRADITIONAL BAMBOO INDUSTRY DEVELOPMENT (1978–1989).

The transformation from tradition to industry

From the late 1970s to the late 1980s, China started to implement the reform and opening up policy. In the late 1970s, Anji's villages were still very poor with an annual per capita income of less than US$50 (250 CNY). A way to change this situation was to develop the bamboo industry; therefore, bamboo cultivation and improvement of the quality of the standing stock became priorities in this period. Bamboo resource management technologies, such as 'Low Yielding Moso Bamboo Forest Rehabilitation Technology' and 'Moso cultivation technologies for both shoot and culm purposes' were introduced and extended, and farmers were encouraged to participate in the research on and extension of high-yielding and high-efficient bamboo forest management technologies.

During this period, the user rights policy of the bamboo forests was also reformed. The 90% of bamboo forests that used to be managed by collectives – communes or production brigades – were contracted to private households on a term of 15 years, which has greatly inspired the initiatives of bamboo farmers. At the same time, the local government had attached great importance to the development of high-yielding and highly efficient bamboo forest demonstrations. Some 40 million CNY was invested by the government as a result of the co-efforts of the government and the communities. The communities input more than 1 million working days of labour, and 4000 ha of high-yielding bamboo

Table 3.3. Bamboo industry development statistics over the period from 1978 to 2013 in Anji, Zhejiang Province, China (data from Zhu Zhaohua and Chen Jianyin, 2013).

Year and rate of increase)/ Industry sector	1978	1988	1994	1998	2002	2005	2010	2013	Increased times over the whole period
Area of managed bamboo forests (ha)	54,822	55,330	56,514	63,330	66,667	67,333	72,400	72,400	0.32
Area of Moso bamboo[a] (ha)	43,200	43,600	44,533	49,867	50,733	52,000	55,287	57,333	0.33
Number of bamboo processing enterprises	30	178	490	1620	865	1880	5300	2160	72.0
Moso bamboo standing stock (millions)	95	110	115	131	135	140	170	170	0.79
Annual culm production (millions)	9.33	12.19	16.00	18.50	20.00	22.00	23.00	28.00	3.0
Production value of bamboo raw materials (millions CNY)	20	160	240	320	420	450	750	770	38.5
Production value of bamboo processed products (billions CNY)	0.003	0.026	0.49	1.70	3.5	4.86	11.00	12.1	4033
Yearly production value of Moso bamboo/ha (CNY)	2700	3000	4500	5700	7500	10,500	10,416	11,188	4.14
Utilization ratio of individual culms (%)	25–30	25–40	30–50	40–60	70–85	75–90	80–90	85–90	3.4–3.6

[a]*Phyllostachys heterocycla var. pubescens.*

forests were established for both culm and shoot purposes, with 15,733 ha of low-yielding bamboo forests rehabilitated, and 5333 ha of new bamboo plantations established. The annual production of commercial bamboo was increased from 7.5 million culms in 1970 to 13.1 million culms in 1989, a rate of increase of 80%; and the annual fresh shoot production was increased from 10,000 to 20,000 tons. With this rapid growth of both culm and shoot production, the local processing industry – mostly manual or semi-mechanized – developed fast. Two bamboo shoot processing enterprises were also established in Anji, and the total number of processing enterprises increased from 30 in 1978 to 178 in 1988.

STAGE II. IMPROVE INDUSTRIALIZED PROCESSING CAPACITY, INCREASE ECONOMIC BENEFITS AND STRENGTHEN MANAGEMENT OF BAMBOO FOREST THROUGH SCIENTIFIC RESEARCH (1990–1999)

> *The expansion of the processing industries, and the upgrading of the sectoral economic efficiency*

After bamboo resource development, the processing industry and value-added products became the priority in this period. Led by enterprises introduced from Japan, China's Taiwan and Hong Kong, and the USA, and with advanced processing machines introduced from Taiwan, a large number of local farmers began to take part in the mechanized processing of bamboo. The number of processing enterprises were also increased from 178 in late 1988 to 1620 in 1998. Some famers started to manage businesses for machinery manufacturing, designing and manufacturing complete sets of processing machines independently. The new machines had become a key driving engine of the fast development of the processing industry.

For example, in Jiquan village, Anji, 60% of the 530 households undertook bamboo mat processing, and over 70 micro-mat processing factories were located in the village. In 1996, this single village produced more than 20 million mats. The village factories were equipped with all the types of machinery necessary for bamboo mat processing, including the weaver, the plastic edging machine, the gauze presser and bamboo slicing machines. as well as mat pressers. The

village realized complete mechanization of mat processing, and the exquisite mat products were all exported to other countries. Jiquan village was well recognized for its professional mat manufacturing.

In order to further push forward the industrialization of the bamboo sector, the local government identified a series of preferential policies to attract investors from overseas, and from Hong Kong and Taiwan, including preferential land, water and power supply and exemption from sales tax for the first 3 years. Professional bamboo raw material and product sales markets were also established and technical training for household/cottage-based factories strengthened. Promotional measures were adopted in product export and sales, and the processing industry of Anji grew rapidly. The production value was increased from 26 million CNY in 1988 to 1.7 billion CNY in 1998.

At the same time as the consumption of the bamboo raw material was increasing, a number of high-yielding and highly efficient bamboo forest management demonstration sites were established under the direct leadership of officials at provincial, county and township levels, and with the joint efforts of technicians and farmers. An 80 ha demonstration site for low-yielding bamboo forest rehabilitation was established by the Director General of Zhejiang's Provincial Department of Forestry, and another 80 ha of high-yielding forests were established by the County Mayor. In addition, 12 villages and 100 professional households were recognized for demonstration purposes, and competition campaigns and projects were carried out to achieve the goals of high-yielding and highly efficient forest management. For example, the '111 High-efficiency Bamboo Forest Project' was implemented to facilitate each bamboo producing village to have at least ten professional households; each of these ten households were to establish a 10 mu (i.e. 0.67 ha, as 1 ha = 15 mu) demonstration site and the annual income/mu from this should not be less than 1000 CNY. These actions fully played the role of advocating the concept of the scientific cultivation, management and processing of bamboo.

To encourage the improvement of bamboo forest management, the local government had allowed a subsidy of 1200–1500 CNY/ha for new plantations, 450 CNY/ha for the rehabilitation of low-yielding bamboo forests, 750 CNY/ha for accessing the new technologies of high-efficiency management of bamboo forest for shoot and culm purposes, and another 750 CNY/ha for newly established demonstration forests. From 1990, the user rights of the state-owned bamboo forests (1.4% of the total) were contracted to individual workers in the state forest farm; as a result, 97% of the bamboo forests in Anji were managed by contracted households.

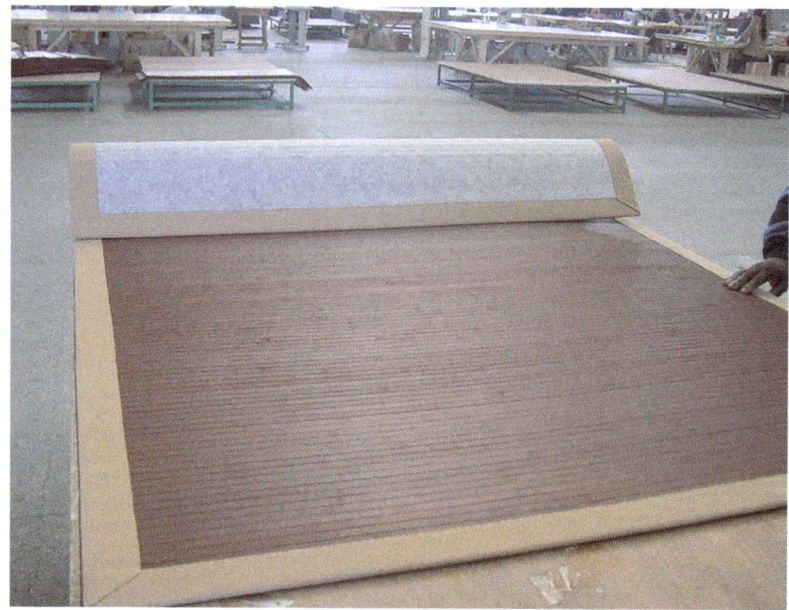

Figs. 3.69–3.72. Bamboo carpet (**Fig. 3.69**), curtain (**Figs 3.70, 3.71**) and mat (**Fig. 3.72**) products in Xiaquan village, Anji, Zhejiang Province, China (Zhu Zhaohua).

Figs. 3.69–3.72. Continued.

The local government prioritized support for research on and the extension of technologies for bamboo cultivation and management, as well as new products and new processing techniques. From 1996 to 2000, more than 10 million CNY was invested in research and 30 million CNY was invested for technology development, adaptation and services. Enterprises were allowed to use pre-tax incomes to cover technical innovation expenses. Large enterprises were encouraged to establish raw material supply bases under the development model of 'leading enterprises + demonstration sites + communities', which realized a co-winning situation among enterprises and communities.

STAGE III. ACHIEVING SUSTAINABILITY THROUGH INTEGRATED ECONOMY AND ECOLOGICAL DEVELOPMENT OF THE ECONOMY (2000–2009)

> *Efforts are made to create a win–win situation between the development of the local economy and environmental conservation, and developing an environmentally friendly economy.*

In this period, Anji's bamboo sector had the following features:

* The processing enterprises gradually grew from household/cottage based microenterprises to medium and large-sized enterprises, with a transfer of management modes from traditional to modern. A large number of medium-sized enterprises came into being, with an average annual production value of 100 million CNY. A certain number of these enterprises even achieved an annual production value of 500 million CNY. By 2009, the total production value of the bamboo sector in Anji reached 12 billion CNY.

* With the scaled-up processing capacity, the local bamboo resources could no longer meet the high demands for raw materials, and the bamboo producers and processors of Anji started to seek raw materials or semi-products from other locations out of the county. They also established new bamboo forests and set up primary processing factories in the bamboo producing areas outside Anji (Chen Jianyin and Yan Guoqin, 2002).

    For example, in 2004, Anji's production of Moso bamboo was 21 million culms, yet the consumption of the processing industry was 55 million culms; thus, 62% of the raw materials came from other bamboo supplying areas. Besides importing fresh culms from surrounding counties, a large number of semi-products were transported from further locations in other provinces. These semi-products were mostly provided by Anji people who had set up plantations and primary processing factories out of Anji. In this way, the bamboo industry of Anji expanded to the whole country. Up to 2004, there were over 50 companies and

more than 3000 people of Anji conducting bamboo industry development in Jiangxi, Fujian, Hunan, Anhui and Yunnan provinces, and the total area of newly established bamboo raw material supply bases reached 37,300 ha.

* The comprehensive development of the bamboo sector in Anji had been a driving engine of the other associated sectors. In order to facilitate the healthy development of the bamboo sector, the local government and bamboo enterprises paid attention to the developments of other associated sectors, such as foreign trade, bamboo product transportation, machinery manufacturing, the packaging industry, construction of factories, forest road construction, chemical industry (e.g. glues and fertilizers particularly for bamboo forests), hardware parts, power supply and telecommunications, etc. Taking the year 2004 as an example again, Anji's total bamboo sector production value was 4.62 billion CNY, and the value of the other sectors driven by the bamboo sector was 1.8 billion CNY. The growth of these associated sectors provided an improved industrial environment for the further sustainability of the bamboo sector. All supplies and facilities necessary for bamboo processing and bamboo forest management could be found in Anji and produced in Anji. So Anji gradually grew into a worldwide supplier of machinery and materials for bamboo industry development.

* The bamboo ecological and cultural industry became the new growth point in this period. The government saw the great potential benefits from the multifunctions of the bamboo forests, and began to develop ecotourism and farm staying in 1999. It established the Anji Bamboo National Forest Park, the China Bamboo Museum and several filming sites. Because of the great significance of rural ecotourism, the government of Anji paid serious attention to the development of farm-stay hotels. A 'Bamboo Farm-stay Service Center' was established; its major services included identifying standards for local farm-stay services and facilities, guaranteeing the service quality; providing market information services and establishing a specialized

website for Anji Farm Stay. Other services provided by the Service Center included strengthening road construction, guaranteeing that every village has access to asphalt pavements and forestry roads, and establishing specialized agencies for guiding and transporting customers, etc. These measures have all helped the healthy and fast growth of the Anji bamboo farm-stay sector (Chen Jianyin and Xuan Taotao, 2007). The ecological and cultural industry became as important as bamboo processing and played greater and greater roles in the development of the local economy.

- From the aspect of forest management, the targets shifted from simply pursuing high economic benefits to the application of comprehensive sustainable management modes. Based on the existing demonstration bases established by villages and professional household producers, a large-scale

modern technology demonstration park was established, the total area of which reached 6700 ha.

According to the new development requirements of the bamboo sector in this new stage, the Anji government implemented the following policies and measures to further lead and support the healthy development of the sector:

- Extension of the user-right contracts: in 2003, the contract period was extended from 15 years to 30 years.
- Tax relief: taxes and fees exempted included the agricultural special duty in 2000, the resource compensation fees in 2004, agriculture funding fees in 2007, etc. In 2003, the taxes and fees for bamboo raw material and product transportation were removed (the 'Green Channel' Project). These measures largely reduced the financial burden on bamboo producers and the processing

Fig. 3.73. A sprayer in Moso bamboo (*Phyllostachys heterocycla* var. *pubescens*) forest in Anji, Zhejiang Province, China (Anji Forestry Bureau).

Fig. 3.74. Applying fertilizer in Moso bamboo (*Phyllostachys heterocycla* var. *pubescens*) forest in Anji, Zhejiang Province, China (Zhu Zhaohua).

enterprises. To date, all taxes and fees related to bamboo forest management have been exempted.

- Diversified management for different types of forests: for example, the bamboo forests on steep slopes were identified as ecological and public welfare forests, and the management of these forests is much less intensified, with vegetation under the bamboo canopy protected. Up to 2002, there were 36,667 ha of ecological and public welfare bamboo forests in Anji, which accounted for 55% of the total. The government issued regulations on harvesting intensity, land preparation methods and standing density. Households managing the ecological and public welfare forests could claim an ecological compensation fee of 390 CNY/ha.
- Forest road construction: from 2006 to 2011, 2300 km of forest road was built in order to facilitate forest management, raw material transportation and ecotourism and farm stay. The government subsidy for road construction was 3500–5000 CNY/km.
- To encourage scientific management of the bamboo forest: the government provided a subsidy of 1.5 million CNY for each modern bamboo forest management technology demonstration park established on a scale above 6700 ha and with access to all the technical standards; and 100,000 CNY for those demonstration forests that were over 130 ha in area. (The other financial support policies remain effective.)
- A special office in the local government: this was set up to provide administrative monitoring and regulatory services for the ecotourism and farm-stay businesses in rural areas, and local management regulations were identified.
- Based on the two existing bamboo product markets (domestic and international), the Anji International Bamboo Art Business and Trade Center was established in Huzhou as a result of public–private partnership (PPP).

Fig. 3.75. A solar-powered black light lamp emitting UVA light placed in a bamboo forest in Anji, Zhejiang Province, China, for insect control/trapping; one lamp can control about 6 ha (Anji Forestry Bureau).

The Center became the largest of its kind in China. To date, over 10,000 rural inhabitants of Anji were engaged in bamboo product sales. The annual per capita income of this group of people was between 60,000 to 300,000 CNY.

- Technology innovation and new product development were still a priority for policy support in this period. Special attention was attached to large and leading enterprises, with an increased strength and scale of financial support.

STAGE IV. THE BAMBOO SECTOR CHANGED FROM SIMPLY BIOMASS PROCESSING TO THE HARMONIOUS DEVELOPMENT OF THE PRIMARY, SECONDARY AND TERTIARY INDUSTRIES (2010–PRESENT)

The integrated and harmonious development of the primary, secondary and tertiary industries have fully explored the multifunctions of the bamboo forests.

After 32 years of development since reform and opening up, the bamboo sector of Anji had gone through a complicated development course from small scale to large scale, from simple biomass utilization to deep exploration of the ecological and cultural values. By 2015, the total annual production value of Anji's bamboo processing industry has reached 13 billion CNY; the industry produced 3000 types of products in nine categories; and the product structure was increasingly improved. The sector was already providing employment for 50,000 people, and the bamboo farmers' income and livelihood were largely improved: the per capita GDP reached 66,739 CNY (US$10,593) and the annual per capita income reached 17,617 CNY, indicating that Anji's rural area had entered the development stage of the 'well-off society'.[1] These figures clearly indicate that Anji's bamboo sector had made great achievements, and had entered into a matured, high-level development stage. The support and

Fig. 3.76a,b. Photos of bamboo-lined forest road in Anji, Zhejiang Province, China (Zhu Zhaohua).

guidance of government policies, which protected and met the benefits of various stakeholders, were indispensable conditions for these
results to have been achieved.

However, all of this does not mean that Anji's bamboo sector was perfect, as there were always new situations, new conflicts and challenges
on the way to development. If these new conflicts

Fig. 3.77. Anji International Bamboo Art Business and Trade Center, Huzhou, Zhejiang Province, China (Anji Forestry Bureau).

and challenges were not properly resolved, the bamboo sector of Anji would not be able to move further forward. The conflicts and challenges were mainly in the following:

1. The fast-developing gross domestic product (GDP) and urbanization process had provided mass employment and income generation opportunities for rural young adults, a large number of whom then shifted their interests from bamboo to other sectors. There resulted in a serious lack of labour in rural areas, and the problem even affected some of the bamboo processing enterprises.

2. Because of the lack of labour, a large number of foreign labourers were hired. In 2012,

over 20,000 foreign labourers were employed by Anji's bamboo industry, and the labour costs rose quickly. In 2012, one temporary worker for bamboo harvesting cost 300–350 CNY/day. Bamboo harvesting is very labour intensive and does not use machinery, so everything had to be done manually, from cutting to carrying the culms to the roadside. This is only cost-efficient when local labour is available; it becomes cost inefficient when labour has to be paid for. So, although the market price of bamboo raw material in Anji was the highest in the whole country, at around 820 CNY/ton, the bamboo farmers could only make about 300 CNY/ton profit after they had paid for the labour that they needed. This was one of the

Fig. 3.78. Moso bamboo (*Phyllostachys heterocycla* var. *pubescens*) harvesting (Zhu Zhaohua).

factors that discouraged the farmers from taking up bamboo forest management.

3. Along with the lack and higher cost of labour, another challenge the processing enterprises had to deal with was the insufficient supply and rising prices of raw materials. In 2010, Anji produced 21 million Moso bamboo culms, while the consumption was 150 million. So the self-sufficiency rate was only 14%.

4. With the increased and higher standards and requirements for corporate responsibility for the environment, the processing enterprises had to increase investment to deal with waste gases and water treatment, thus lowering carbon emission and labour insurance. These measures resulted in a significant increase in the cost of products.

5. With the rapid development of the other industries in Anji, although the bamboo sector was still developing fast, the importance of the sector in the total GDP of Anji was decreasing, as was the dependence of the rural economy on the bamboo sector. The micro- and small-scale, sparse and separate management of bamboo forests were considered not to be compatible with modern management modes.

In the face of these new challenges, the local government identified a series of new policies and measures to push the bamboo industry forward.

Further stabilize land user rights policy, allow heritage, gift, transfer and mortgaging of the user rights. To motivate farmers to take up bamboo forest management, especially to encourage the participation of cooperatives and enterprises, the local government extended the land user rights contract period from 30 to 50 years. The government also allowed the transfer, gift and heritage of the bamboo forest user rights, so as to allow micro-and small-scale forests to be transferred to large-scale managers, cooperatives or enterprises. The management body was shifted from individuals to shareholders, while the user rights of the bamboo forests could be transferred or mortgaged for loans. These measures did not only protect farmers' interests, but also provided more options for the farmers to take part in bamboo forest management. Large-scale professional management households, cooperatives and enterprises were also motivated to join the management of bamboo forests,

Fig. 3.79. A forest user rights certificate issued by the government in Anji, Zheijiang Province, China (Anji Forestry Bureau).

and they have become the main management body that is working in the field professionally and actively.

From 2009, bamboo farmers were guided to transfer the forest user rights as shares, so as to activate and shift forest management from private individuals to legal persons. The management method was changed from dispersed, low-efficiency management to intensified, high-efficiency management. Resource allocation was activated, and static capital was changed into dynamic capital. By 2014, 30 shareholding cooperatives were established in Anji, the total value of the user rights was 290 million CNY, and the total area managed in this way was 4760 ha; some 3673 households became members of the cooperatives and their average annual increase in household income was 11,000 CNY. Through establishing the shareholding cooperatives, allowing transfer and mortgage of the bamboo forest user rights, and the construction of demonstration parks, the key elements of modern forestry were fully realized (Anji Forestry Bureau, 2013b).

The changes that are described here were Anji's new exploration in recent years under the reform and opening up mandate in order to keep pace with the new situations of local socio-economic development. The changes were positive in development, but in their implementation, there were always new conflicts and problems.

Construct beautiful countryside. Since 2008, the Anji government set the goal of 'establishing beautiful villages, every household engaged in new means of livelihood, achieving harmony everywhere and all people live in happiness' and the project of 'improved environment, industry, service and personal quality'. By 2012, the government had established 164 special villages, each with its own unique products, attractions and scenery. Anji was rated as the first national ecovillage and eco-civilization demonstration site. Besides this, the Anji government put in great efforts to develop key tourism spots and speciality tourism products based on its rich history and culture. With these measures, Anji became a beautiful, comfortable, harmonious and rich county, thus creating conditions for the development of the ecotourism and leisure industry, so that it became one of the best counties to live in in Zhejiang Province and China.

Integrate the primary, secondary and tertiary bamboo industry. The Anji government and the local people learned lessons from a severe snow disaster in 2008, a typhoon in 2012, and heat and drought in 2013, and gradually understood that the county's development model should shift from either complete economy or complete ecology to 'ecological economy and economic ecology'. This involved changes to the primary, secondary and tertiary bamboo industry, which are described below.

Fig. 3.80. A beautiful village surrounded by bamboo forest in Anji, Zhejiang Province, China (Anji Forestry Bureau).

In the *primary bamboo industry*, changes were made from low efficiency and small scale operations to relatively large scale and scientific management. The management intensity and amount of fertilizer used were reduced; the use of organic fertilizer was increased; undergrowth vegetation was protected; and interplanting or livestock raising, such as raising mushrooms and chickens in bamboo forest, was applied. By 2013, there were five professional cooperatives raising chickens, ducks and geese in bamboo forests, and four households producing more than 30,000 fowls. The business scale has been gradually improved, and labouring intensity has been reduced with the application of mechanized operations. The infrastructure in the bamboo forest area has been enhanced. For example, the government increased the subsidy for forest road construction to 35,000–50,000 CNY/km, and by 2013, 150–200 km of forest road had been improved. As some farmers said, the 'demonstration site has become a tourism spot, bamboo farmers have become stockholders and

products have become souvenirs', thanks to the new management concept. This is a great change of concept and development model, and is because the bamboo forests are no longer simply entities that provide biomass, they are also resources that people can tap for beautiful landscapes and scenery, fresh air and pure water, as well as a comfortable environment for tourists to stay in. 11 tourism scenic spots for bamboo forests had been established, and bamboo science parks with an area of 13,867 ha had been developed. In 2005, Xi Jinping, who was then the Secretary General of the Party of Zhejiang Province, visited the Yu Village of Anji, when seeing the hills and hills of bamboo forests, he made a famous statement to express his heartfelt thoughts: "Lucid water and green mountains are invaluable assets." This famous statement now has become China's guiding principle for ecological civilization construction.

In the *secondary bamboo industry*, the government established industrial development zones to speed up enterprise upgrading and

Fig. 3.81. A leisure site in bamboo scenery in Anji, Zhejiang Province, China (Zhu Zhaohua).

Fig. 3.82. Poultry under a bamboo forest in Anji, Zhejiang Province, China (Zhu Zhaohua).

product innovation and develop branded products. The government also encouraged enterprises, especially large-scale enterprises, to produce standardized and scaled branded products with high quality. In 2013, there were 2162 bamboo product and equipment enterprises in Anji. Among them, two are leading enterprises in China, 31 are provincial leading enterprises in agriculture and forestry, 59 are large-scale enterprises, and 11 have a production value of over 1 billion CNY and 29 a production value of over 50 million CNY. Flooring production has exceeded 50% of the world's total. Bamboo mat industry production reached 6.14 billion CNY, which was 70% of the whole market in China, and machine manufacturing accounted for 80% of the market in China. A complete processing chain from raw material to finished products was established and the industrial utilization rate exceeded 90%. Anji realized almost full use of every part of bamboos. Product structure shifted towards high added-value products. For example, outdoor flooring, decoration, veneer. Bamboo construction and furniture are becoming new fashions. In order to improve product quality and develop self-owned brands, with the government's encouragement, enterprises actively participated in the drafting of standards – and some large-scale enterprises even got involved in the setting of national and provincial standards. To cut the cost of labouring and raw materials, many enterprises are undergoing overall technical updates to promote the automation of manufacturing, to speed up efficiency in terms of time.

In the *tertiary industry*, the changes that have been made are of three types: market development; development of the ecology and cultural industries; and development of carbon trade.

- (a) Market development. Anji has rich experience in domestic and international market development, and its bamboo products have been exported to Europe, America and Japan, and now are entering the Asian, Latin American and African markets. People find it hard to believe that most bamboo enterprises were originally small and private and that the owners mostly came from rural areas and had no high-education background; that they were among the first group of 'farmer entrepreneurs' after the opening up and reform policy of the country; that they started from homestead businesses and developed them into modern, large-scale enterprises; and that they were not only good at management, but also did very well in developing both domestic and international markets. These developments indicated that the provision of an adequate policy and investment environment could inspire people's initiatives and creativity.

- (b) The development of the ecological and cultural industries has put the multiple functions of the bamboo forests into full swing. The number of tourists to Anji in 2015 reached 14.9521 million, which was 32.5 times the number of the local residents (460,000). The tourism production value in the same year was 17.564 billion CNY, of which 7.02 billion (40%) was bamboo related. The elevation of the level of ecotourism development was realized through a series of programmes, including: the construction of multifunctional agricultural zones such as leisure farms, tea gardens and bamboo forest demonstration sites, and the application of a bunch of comprehensive, thematic and high-end tourism projects, such as astronomical observations, panda and bamboo forests and ecological filming.

These programmes upgraded ecotourism in Anji to a higher level and promoted the development of farm stay. Culture spots with a rich history and cultural heritage were also developed. Based on the local bamboo resources, the Anji bamboo ecotourism industry started to transform into a leisure and health preservation industry, and along with this transformation, model villages and households were established, and new and higher standards were developed for farm-stay hotels. These high-quality touring sites were constructed by combining the scientific management of the bamboo forests, bamboo product processing and the ecological and cultural industries. A number of five-star hotels with bamboo structures were established in the county.

Fig. 3.83a,b,c. Photos of a pressing machine for processing pressed bamboo material in Anji, Zhejiang Province, China (Zhu Zhaohua).

Fig. 3.84. Bamboo lumber after cold pressing (Zhu Zhaohua).

Fig. 3.85. Different types of floorings made from pressed bamboo material (Zhu Zhaohua).

- (c) Developing carbon trade and taking advantage of the strong carbon sink capacity of the bamboo forests. At the Conference of the Parties (COP) 2012 – the 2012 United

Nations Climate Change Conference – Anji, representing the global bamboo producing areas, signed the Framework of Research, Trials and Demonstration on Bamboo

Forest Sustainable Management, Bamboo Product Carbon Storage Calculation and Trade （竹林可持续经营和竹产品储碳计量和交易研究与试验示范区建设框架协议）, in partnership with the China Green Carbon Foundation and Zhejiang Agriculture and Forestry University. Anji, therefore, had become the first ever global bamboo forest carbon sink and product carbon storage trial and demonstration zone in the world. So now, systematic research on the mechanism(s) by which Moso bamboo forest acts as a carbon sink, and auto remote sensing observations have been carried out in Anji. As climate change has become a global concern, bamboo forests will play an increasingly important role in carbon trade. The research that is being done in Anji will provide a solid basis for future bamboo carbon trade, and will become a new growth point for the bamboo sector.

The above accounts of the primary, secondary and tertiary bamboo industries of Anji's bamboo sector in Stage IV of its development indicate that they have integrated into a cluster and entered into a comprehensive development period. In the face of a large and complicated cluster of industries, the local government needs to consider policies to take care of the benefits of various stakeholders, inspire initiatives and facilitate by providing access to both software and hardware in order to achieve a higher level of development of the bamboo sector and a multi-win situation. This fourth stage in the development of Anji's bamboo industry (i.e. 2010 to the present) is an extremely hard and transforming stage, while at the same time it opens up a wide new range of development possibilities for the bamboo sector. There has been such a huge change that the bamboo sector of Anji is no longer the one we used to understand – just bamboo forest management and bamboo processing. This new comprehensive bamboo sector will need a comparatively longer period to mature, and it will finally need to be integrated with other types of industries and sectors.

Through all the Stages I–IV described in Section 3.5.2, different policy compositions and systems were developed to meet the needs and situations in different stages, while guiding and supporting the bamboo sector towards sustainability. A precondition of all these was the thorough investigation and understanding of the actual needs of various stakeholders, and of the zeal and sense of responsibility involved. A responsible and healthy policy environment should always be more important than cash.

For different countries and regions, the challenges and problems can be different, and so will be the policies and measures to be taken. The goals will be balanced benefits among stakeholders – multi-win, continuous technical innovation and the upscaling of industry to meet market demands. One thing that has to be kept in mind is that the farmers and the enterprises are the principal bodies in the bamboo sector and there should be no interference with their decision making and business management processes. A proactive government should only provide services and guidance, and should know its responsibilities.

3.6 Personnel, Institutional Capacity Building and Multi-participation

The bamboo sector is a complicated sector that covers a wide range of fields and various groups of people. The multi-participatory nature of the sector and the support that it receives on various aspects are the indispensable conditions for its survival. It is also extremely important for all stakeholders to have a deep knowledge of bamboo, and be aware of its significance for poverty alleviation, rural economy development, and ecological and environmental protection. Only when they understand the bamboo industry well can they take an active part in it.

Awareness raising then, is important, and awareness raising and motivation are the basis of personnel and institutional capacity building. Capacity building can take various forms, and besides various types of training, could also be organizing different institutes and experts to provide consultancies, or improving management models.

3.6.1 Awareness raising

Awareness raising for different groups of people – the MOST international training programme

First of all, the prioritized target groups for the training programme need to be defined. The usual prioritized groups were bamboo farmers and factory workers. However, a large amount of practical experience from the programme has proved that the key institutions and key individuals should be prioritized first, such as the relevant officials in government departments, the investors and entrepreneurs, and the relevant researchers and experts. Only when these key individuals have a clear understanding of the bamboo sector will they be able to determine whether or not to make bamboo an option and how to do this. In order to build the capacity of these key individuals and institutions, a very finely designed programme needs to be prepared.

From 1997 (when INBAR was founded) to 2015, sponsored by the China Ministry of Science and Technology (MOST), the authors have organized 17 international training workshops in China, in partnership with the International Farm Forestry Training Center (INFORTRACE) of the Chinese Academy of Forestry and the Zhejiang Lin'an Modern Forestry Science and Technology Service Center and INBAR. Some 595 participants from 58 countries have been trained in these workshops. As a follow-up service, we have also conducted 24 workshops for a number of countries and regions at the requests of alumni and partners; these countries include Chile, the Philippines, Myanmar, Vietnam, Laos, Thailand, India, Brazil, Colombia, Cameroon, Ghana and Kenya, and the participants trained from these countries numbered 430. Most of these participants were senior officials in government, experts and enterprise leaders. The training programme in China greatly increased their knowledge on bamboo and its industrial development. According to feedback from alumni, some of them organized training courses back home, established NGOs (non-governmental organizations) for bamboo development, wrote articles or technical reports and published them, established new bamboo enterprises, etc., so these activities have benefited even more people around the world. Here are two examples that demonstrate the impact of the training programme.

CASE STUDY 1: ROMUALDO L. STA ANA – AN ALUMNUS FROM THE PHILIPPINES. Romualdo Sta. Ana from the Philippines participated in the MOST/INBAR training programme in 1999 and then came back to this same programme eight times from 1999 to 2015. He was a founding member of the Philippine Bamboo Society, which was originated in 1997, and registered in 1998 as the Philippine Bamboo Foundation.

He organized more than 30 training workshops in the Philippines with more than 900 participants (see Sta. Ana, 2015). The trainees included officers of government agencies and private companies and organizations/NGOs as well as some retired officers of the Philippine military. One of the retired colonels from the Armed Forces of the Philippines has already developed a 500 ha bamboo plantation after attending the training. Romualdo Sta. Ana introduced many of the technologies and the best practices learned from the workshops attended in China, and also from other training on primary processing and bamboo furniture making. He also conducted training in bamboo furniture making in other countries, including Nepal, Nigeria and Peru.

Leading study tour delegations to China. Romualdo Sta. Ana organized nine delegations, in total 67 people, to participate in the MOST/INBAR training programme or conduct special study tours on bamboo in China in 2004, 2006, 2008, 2009, 2010, 2011 and 2013. The delegation members include three Governors, a number of Mayors and local government officials, entrepreneurs and scientists. These delegations pushed forward the development of the Philippine's bamboo sector after their return. Below are some of the stories told by Romualdo Sta. Ana:

1. Armando Mendiola Jr and Alfredo Rabena attended the International Training Workshop on 6–20 September 2004 with me. Armando Mendiola Jr makes bamboo-veneered furniture which is exported to other countries.
2. Buenafrido Berris, Mayor of the Local Government Unit Calauan, Laguna, Philippines: after the visit, Mayor Berris implemented bamboo projects

Fig. 3.86. A Chinese expert training local Ghanaian technicians (Zhu Zhaohua).

Fig. 3.87. Romualdo Sta. Ana visiting China (Zhu Zhaohua).

in Calauan and established a bamboo factory for primary processing and manufacturing chopsticks and barbeque sticks.

3. Proceso J. Alcala, Congressman, House of Representatives, Philippines: Congressman Alcala asked me to conduct a training workshop on Bamboo Stand Rehabilitation and Nursery Establishment and Management and implemented several bamboo projects. He even invited Prof. Zhu and Prof. Wang Anguo to visit his bamboo projects in the province of Quezon. Mr Alcala is now the Secretary of the Department of Agriculture and has included bamboo in the projects of his department.

4. Edgardo C. Manda, Manager of the Laguna Lake Development Authority (LLDA) and Presidential Assistant for Region IV Office of the President of the Philippines: Mr. Manda hired me as consultant for the bamboo projects that he adopted for the LLDA. This included training and the establishment and management of bamboo nurseries and plantations to protect the environment around the lake, which is the biggest one in the country.

5. Rene E. Cristobal, Chairman/CEO, Innovative Concrete Elements, Inc., Silang, Cavite, Philippines: Mr Cristobal and I organized a company together, Bambusapinas, Inc., which is setting up a model bamboo project in La Union Province in the northern part of the country. Our project will include nurseries and plantations, and a training and processing centre which will include primary processing, furniture making, bamboo shoot processing and laminated bamboo products.

6. Abigail Chon, from Calauan, Laguna, Philippines, learned a lot from the lecture–presentations and field visits, especially about bamboo development in China, the full utilization of bamboos and the value chain.

7. Mr Bituin and his family make solid and bamboo veneered furniture which are exported mostly to the USA, Europe and the Middle East.

8. Messrs Calleja, Lagarto, Cruz and Badana have acquired bamboo processing machinery from China and established a factory that includes primary processing and the making of barbeque sticks, chopsticks, toothpicks and bamboo floorboards in Iloilo.

9. Miss Soriano owns a 20 ha property near Laguna de Bay, Philippines, where she worked with me and established a giant bamboo nursery and plantation to support the LLDA bamboo programme.

Consultancies for government, the private sector and other countries. Romualdo Sta. Ana carried out consultancies for the following organizations/countries:

1. Provincial Government of Tarlac. He helped the Government of Tarlac in conducting a bamboo resource survey. He also provided several training sessions on bamboo forest rehabilitation, harvesting and management, and conducted training on nursery establishment and management and primary processing.

2. Laguna Lake Development Authority. He was hired as consultant for the bamboo development project started by Mr Edgardo Manda after his visit to China in 2006. The main purpose of the project was erosion and pollution control.

3. Philippines Department of Trade and Industry (DTI). Although he was not hired as a consultant by the DTI, they invited him to many of their fora where he made presentations about bamboo for environmental protection and livelihood. He told the forum participants that his dream is to make the Philippines second only to China in the export of bamboo products. The DTI has adopted this goal and has implemented many programmes together with other government agencies and the private sector. The Secretary of the DTI has allocated PhP (Philippine Pesos) 800 million for 700 Shared Service Facilities (more commonly known as Common Service Facilities) throughout the country. The money is intended mainly for the purchase of machinery and equipment that the DTI donates to local government units. This will support the government programme that requires that 25% of school desks must be made of bamboo (the annual requirement for desks by the Department of Education is valued at PhP 1 billion).

4. MFI Foundation Inc. The MFI Foundation owns a 63 ha property in Jalajala, in Rizal Province, Philippines where it operates a farm school. Romualdo Sta. Ana conducted several training sessions on bamboo stand rehabilitation and nursery establishment, which were attended by participants from all over the Philippines, including retired military personnel and government employees as well as private landowners, farmers and students of the farm school. He also helped the school to establish a bamboo nursery and planted *D. asper* in the property.

5. Planters Products, Inc. (PPI). He has worked as a consultant for the company, and his job included the conduct of training on bamboo nursery establishment and management in areas where PPI buys bamboo poles, improving the layout of their existing bamboo chopsticks and barbeque sticks, and training the supervisors and workers.

6. Rotary Club of Makati San Lorenzo, Philippines. He was the consultant for the bamboo development project funded by Rotary International to help the indigenous people living in the Sierra Madre Mountains.

7. Short-term consultancy for Peru. Romualdo used to be one of the consultants of the International Tropical Timber Organization (ITTO)-funded bamboo project Peru Bambu.

Publications and training materials. Romualdo Sta. Ana prepared English and bilingual manuals for the workshops that he conducted.

The manuals he used in the Philippines are in English and Tagalog (Philippine dialect) with a lot of pictures. He wrote the manual with the English paragraph on the left of the first column and the Tagalog translation on the right-hand column directly opposite the English version. He usually sent manuals with the English version and blank sections opposite it in advance to the countries where he would be conducting the training so that they were translated into Spanish for his training in Peru and into Nepalese for the training in Nepal.

CASE STUDY 2: IMPACT OF TRAINING PROGRAMME ON THE DEVELOPMENT OF VIETNAM'S BAMBOO INDUSTRY. Since the establishment of INBAR, there have been 20 participants from Vietnam taking part in the MOST/INBAR training programme. Most of them are from universities and research institutes. In 1999, Vietnam became a member country of INBAR, and the Vietnamese Ministry of Agriculture and its main bamboo producing provinces started to attach importance to bamboo

Fig. 3.88. A fish farm in sea surrounded by bamboo in the Philippines (Zhu Zhaohua).

Fig. 3.89. A bamboo hotel on the beach by the sea in the Philippines (Zhu Zhaohua).

Fig. 3.90. A bamboo (*Bambusa blumeana*) culm carving (Zhu Zhaohua).

development. With the support of international organizations such as the Prosperity Initiative (PI, a UK-registered international organization set up to help develop the bamboo sector for the benefit of business, farmers and others in the industry in partnership with government partners in the region; see PI, 2010), Oxfam, Winrock International (which works in US and international development with a focus on social, agricultural and environmental issues), SNV (the Netherlands Development Organization), GRET (a French international NGO) and USAID (the US Agency for International Development), Vietnam sent seven delegations, more than 140 people from 2006 to 2011 to China to study bamboo industrial development experiences. The composition of these delegations was quite representative, with each group usually having policy makers, technicians and leading entrepreneurs and officers from international organizations. For example, in 2007, a 38 member delegation from Vietnam came to China, that included eight Director Generals from the Forestry Departments of four major bamboo producing provinces, ten district mayors, four officers from international organizations, eight entrepreneurs and eight scientists.

In 2010, a 32 member delegation led by the Vice Minister of Agriculture visited China for a bamboo study tour. The group was sponsored by Winrock International. After this tour, the Vietnamese government fully realized the importance of policies on bamboo industry development. In the same year, the Ministry of Agriculture sent a special delegation to China to

Figs. 3.91–3.93. The home of the former First Lady, Imelda Marcos (伊梅尔达 马科斯) of the Philippines in Negros. The inside decoration and furniture are all made of bamboo (Zhu Zhaohua).

Figs. 3.91–3.93. Continued.

conduct a bamboo policy study tour. The delegation had systematically studied China's policies for bamboo industry development, and conducted a workshop to discuss China's policies and how to adapt them for Vietnam. The group also proposed a 5 year plan for Vietnam's bamboo and rattan industry development (Zhu Zhaohua, 2010a,b).

The above-mentioned study tours played an indispensable role in strengthening the confidence of government officials from high levels to grass-root levels, as well as that of entrepreneurs, in bamboo industry development; at the same time, it facilitated their determination. In October 2010, the Vietnamese Ministry of Agriculture organized a National Conference on Bamboo Industry Development Policy. The conference drafted the document 'Decision on Encouragement Policies for the Bamboo and Rattan Sector' and presented it to the Premier.

The above series of actions, study tours to China and a Chinese expert consultancy in Vietnam played an active role in promoting the bamboo sector in Vietnam. By 2010, Vietnam had developed 100,000 ha of new bamboo plantations; the annual production value of the bamboo sector reached US$250 million; annual raw bamboo production reached 5.5 billion culms and shoot production reached 140,000 tons. According to the statistics of the Anji Bamboo and Wood Machinery Association, Vietnam imported over 100 production lines from 2000 to 2012. Now, Vietnam produces a series of bamboo products, such as fine handicrafts, furniture, engineering board, flooring, pressed bamboo materials, various types of mats, sticks (chopsticks, toothpicks, barbecue sticks, incense sticks, etc.), fermented and dried shoots, etc. So Vietnam has become one of the countries that has a comparatively developed and large-scale bamboo industry, and it has been estimated that the bamboo sector of Vietnam may be ranked as third place in the world, after China and India.

THE COMPOSITION OF THE DELEGATIONS ATTENDING TRAINING PROGRAMMES CAN DIRECTLY AFFECT THE RESULTS OF THE TRAINING. Experiences on international training from 1999 to 2015 indicated that the composition of the delegations from a country or a region may well impact the awareness of and future actions on bamboo development in the country or region concerned. If the

Fig. 3.94. The government of Vietnam delegation visiting Lin'an, Zheijiang Province, China (Zhu Zhaohua).

delegation is only composed of technicians and researchers, but no policy makers and entrepreneurs, the impacts may be limited to workshops and publications introducing experiences from China, and new bamboo research projects. It would also be quite hard for scientists to push bamboo industrial development. However, if the delegation is composed of policy makers, entrepreneurs and scientists, this would be an ideal composition, and it would be even more ideal if officers from international organizations were also included, as there would then usually be real actions on bamboo sector development after their return. Some groups have even started making plans during the training programme, and held special meetings with the experts and organizers of the workshop to discuss future plans for bamboo projects.

CONCLUSION. The two case studies above show that it is important to build the capacity of key government officials, entrepreneurs and scientists. Only when these people are aware of the

importance and the proper ways of developing bamboo industry can they provide a favourable environment for the bamboo sector back in their own countries/regions.

Capacity building or training for decision makers and entrepreneurs should be very flexible and effective. The form could be, for example, high-level forums, international seminars, short-term workshops, field trips to successful pilot sites at home or abroad, or lectures or consultancies by relevant experts at home or abroad, etc.

Case study 3: Awareness raising for bamboo farmers and communities in Lin'an and Anji

Training for bamboo farmers and communities must be combined with production practices such as the rehabilitation of low-yielding (natural) forests, plantation development, demonstration site construction and the establishment of homestead factories, etc. However, it is important that awareness raising and training activities are carried out before the improved production practices are put in place. The cases of both Lin'an and Anji have indicated that the inspiration and motivation of bamboo farmers and local communities were the most important part of the mission at each stage of bamboo sector development, because they were the final beneficiaries of bamboo development. Various methods of training were used for farmers and communities:

- Training courses: in the first 5 years of Lin'an's bamboo shoot industry development, bamboo technology training courses trained more than 5000 farmers each year.
- Demonstration households: technical training was first provided to a number of demonstration households and then these households trained the other farmers.
- Construction of the demonstration sites: farmers and community members were invited to participate in the management activities in the demonstration site, in order to receive on-site training in the technologies concerned.
- Bamboo shoot industry newsletters: newsletters popularize knowledge on scientific bamboo plantation management and provide information on markets and successful cases, thus allowing a wider range of local communities to be acknowledged and encouraged to participate in bamboo industry development.
- Media communication: local radio, TV and newspapers were all used to promote the bamboo industry.
- Raising the awareness of the local people: the organization of group activities and allowing opportunities for equal participation are effective methods of awareness raising.

3.6.2 Stakeholders' associations

Linking enterprises, communities and policy makers

The development of the bamboo sector in a country or a region may not be realized simply through individual efforts; rather, it needs multi-participation. The various stakeholders of the bamboo sector should be organized together, establish partnerships, share information, support each other and even become one organization of the same interests. This type of organization already existed in some of the countries and regions that wanted to develop their bamboo sectors – as organizations such as bamboo societies foundations, academies and societies, etc. NGOs like these play important roles in bamboo sector development, by linking and binding the stakeholders together.

Non-governmental organizations and bamboo development in China

There are a number of bamboo-related NGOs in China: at the national level, there is the China Bamboo Industry Association and the Branch Society of Bamboo of the Chinese Society of Forestry. The China Bamboo Industry Association is responsible for proposing the national bamboo industry development plan and policies to the government, and it organizes the China Bamboo Cultural Festival every 2 years. The Branch Society of Bamboo holds regular academic seminars to report and exchange research results. There are presently more than 400 members in the Society, and they are from different parts of China and work in various research institutes, universities and enterprises. The society also has its own journal, the *Journal of Bamboo Research*.

There are also bamboo industry associations and academic societies at the provincial and county levels. The associations at county level are probably the most active, and their services are likely to be much more specific and direct than those of the associations at higher levels. For example, the Anji Bamboo Industry Association provides services for bamboo cultivation, processing and product sales in the whole county, and it advises the local government on strategy and policy making according to members' needs and suggestions. Besides this Association, professional societies have been established according to need, such as the Anji Bamboo and Wood Machinery Society, which was founded by the joint efforts of bamboo and wood machinery manufacturers in Anji, and the Anji Chamber of Commerce for Bamboo Mats, which was founded by 120 bamboo mat producers in Anji in August 2011. The purposes of these professional societies were to provide regulations, help each other in improving technologies, share information on markets and technology, and coordinate sales and marketing activities to avoid chaotic competition. Below is a detailed introduction to the example of Lin'an Bamboo Industry Society.

Case study 4: the Bamboo Industry Society of Lin' an

The Bamboo Industry Society of Lin'an was launched in 1985 by a number of bamboo shoot processing enterprises, bamboo plantation demonstration households, bamboo experts, researchers and technicians, along with some of the retired government officials who had a good understanding of the bamboo sector in Lin'an. Since its establishment, the society has undertaken the activities described below.

INVESTIGATIONS AND SURVEYS OF LIN'AN'S BAMBOO DEVELOPMENT STRATEGY AND POLICY MAKING. According to the needs and situations in the different development stages of the bamboo sector, the Lin'an Bamboo Industry Society has organized its members to carry out investigations and research, and make proposals to the government on planning and policies based on the results. From 1985 to 2005, the Society made more than 30 proposals to the government. A main goal it set for the government was that 'bamboo become a polar industry and facilitate the local economy to achieve two folds'. The society's proposals were accepted by the government, and the first decision on bamboo shoot plantation development was to establish 6700 ha of new plantations in 10 years.

When it was seen that bamboo shoot production brought great economic benefits to the rural people of Lin'an, people started to plant bamboos everywhere without consideration of the site conditions or of any scientific management. For example, the planting of large areas of monoculture can pose a serious threat to the sustainable development of the bamboo sector. At this time, the society proposed the approach of 'Deep development of the bamboo resources and pushing forward the sustainable development of the bamboo industry'; thus, adaptive measures were advised according to the features of the three existing types of bamboo forest. These were to:

- adequately enlarge the areas of Moso bamboo forests for both shoot and culm purposes;

- limit further extension of the areas of *Ph. praecox* and *Ph. vivax*, but promote unit production and apply quality control; and

- for natural stands of *Ph. nuda*, whose shoots are suitable for drying and preserving, rehabilitate and cultivate the forest, and improve its regeneration capacity and shoot quality.

The local government applied these measures according to the society's advice, and special meetings were held with relevant departments of the government, enterprises and communities to discuss the implementation issues of these measures.

NETWORKING AND PARTNERSHIP DEVELOPMENT, PROVIDING FULL SERVICE TO THE BAMBOO SECTOR. The research members of the Bamboo Industry Society of Lin'an have carried out research on practical bamboo technologies, and they have also taken on the responsibilities of extending these technologies. As a result, a series of results from practical research has significantly increased the productivity of bamboo shoot growing. For example, the 'Early Shooting Technology of *Ph. praecox*' made it possible for the spring shoots to emerge 3 months earlier, and as a result, the value of the off-season shoots on the market was increased tenfold, and the highest record of production of off-season fresh shoots was 57 tons/ha. The society has also edited local technology manuals and standardization documents, such as 'Cultivation Technologies of Bamboo Forests for Shoot Purposes', and 'Lin'an Non-polluted Bamboo Shoot Cultivation Technology Standards'. Eleven scientific research and extension projects were organized and implemented by the society during the period from 1985 to 2005, and the results from these projects were recognized and rewarded by the China Ministry of Forestry, the Zhejiang Provincial Government and the Hangzhou Government.

The society also provides great support to key bamboo projects in Lin'an, such as: the 'Construction of the 670 ha Demonstration Sites for *Ph. praecox* and *Ph. vivax* Cultivation'; the 'Construction of 2000 ha of Demonstration Sites for Moso Bamboo Forests for Both Shoot and Culm Purposes'; and the 'Construction of 6700 ha of Demonstration Sites of *Ph. nuda* Forest Management for Dried Shoot Production'.

The society has provided services in personnel cultivation and capacity building as well. Technical advice, training and consultancies were provided for bamboo farmers, enterprises, workers and sales persons. From 1990 to 1995, the society organized 85 lectures and distributed 30,000 posters targeting different types of stakeholders. In this period, 125 issues of the *Bamboo Newsletter* were published, and technical manuals and TV programmes were also produced and distributed. Training courses were provided to the rural communities, and about 15,000 trainees joined these courses. Experts and technicians were organized to provide on-site demonstrations and instruction for local demonstration households. Researchers from relevant institutes and universities were also invited to Lin'an to attend exchange activities, so the society has been keeping pace with the latest developments in the bamboo sector.

EXTENSION OF THE SUPPLY CHAIN.   According to the new development strategy in the second stage, the local government needed not only to prioritize raw material production, but also to make great efforts in opening up new markets. In order to extend the value chain of the bamboo sector, the Bamboo Industry Society of Lin'an started to provide services to the processing enterprises to improve and extend the supply chain. These started based on the strategy of 'developing raw material first and then promoting market sales with major efforts focused on the processing sector'. In 2003 and 2004, the society helped Kangxin Food Co., Ltd., a leading bamboo shoot processing enterprise, to realize contracts with 10,000 professional shoot producing households, a move that solved the raw material supply of the company, while at the same time, facilitating the formation of the new development model 'company + processing + community' (Wang Anguo, 2005).

From the above account of the Bamboo Industry Society of Lin'an, it can be seen that it played significant roles in catalysing cooperation between the government, enterprises and rural communities, as well as capacity building (Wang Anguo, 2009).

Case study 5: introduction to the Colombia Bamboo Society

The Colombia Bamboo Society (CBS) is the earliest bamboo related NGO in Latin America. It was launched by 65 members that included domestic and international experts, engineers, architects, entrepreneurs, craftsman and farmers. It holds regular annual activities and actively engages in consultancy, research, product development, bamboo forest management, giving advice to the government and international cooperation programmes. The activities of the

Fig. 3.95. Members of the Colombia Bamboo Society (Ximena Londoño).

CBS have played a great role in the development of the bamboo sector in Colombia and the Latin American region. A number of the experts in the CBS have collaborated and published a number of highly esteemed publications, which have made a big impact on the world's bamboo research. The following is an introduction to the CBS written by Dr Ximena Londoño, Chairperson of the CBS.

HISTORY. The CBS is an NGO that was born on 12 December 1998, in the municipality of Montenegro, Quindío, Colombia, with the purpose of providing knowledge on the fantastic native bamboo *G. angustifolia* to academics, farmers, the government and businesses at the regional and country level, as well as links between these groups of people. Its headquarters is in the city of Armenia and it has 70 members, both national and international, and from different bamboo-related sectors: researchers, engineers, architects, workers in industry, handicrafts workers, farmers, artists and technicians.

OBJECTIVES. The objectives of the CBS are to assemble our members interested in the Guadua bamboo sector together with universities, autonomous corporations, the government and NGOs, to encourage the planting, study, preservation, transformation and industrialization of the Guadua bamboo in order that it becomes important in the conservation of the environment and for the regional and national economy, and generates benefits at several levels. Other objectives of the CBS are to promote and carry out social projects, develop social promotion plans and provide general training for interested people, with emphasis on underprivileged individuals, and to promote a bamboo research centre, organize events and publish information that contributes to spreading and increasing knowledge of the Guadua bamboo.

ACHIEVEMENTS. During its 15 years of operation, the CBS has contributed significantly to the transformation of the use of Guadua bamboo in Colombia, and has made large contributions to the domestic and worldwide efforts in research on, and industrialization, standardization and promotion of Guadua bamboo, which is considered to be the lumber source of the future. One of the most important achievements of the

CBS was to have Colombia admitted to INBAR in November 1999, during the government of President Andres Pastrana.

During the reconstruction of the coffee-growing region (1999–2001) in Colombia, after the earthquake in 1999, four low-income housing development projects were implemented, and Guadua bamboo houses were established and given to poor women affected by the earthquake. The bamboo houses used designs drawn up by members, and several training workshops were given on seismic resistant building structure.

The CBS has provided support for student theses in diverse universities of the country in order to stimulate research in the area of Guadua bamboo. It has also led and carried out several research projects with the Technological University of Pereira (Colombia) and the National University of Colombia, as well as with Imperial College London and research centres such as Cenicafé (Centro Nacional de Investigaciones de Café, Colombia) and the Energy Research Centre of the Netherlands.

Some of the research that has been supported is outlined below:

- Quantification of the carbon sump effect of *G. angustifolia*. This study found that the species has the potential to fix 33.2 equivalent tons of CO_2 a year with new plantings.
- Anatomical characterization of the culm of *G. angustifolia*. This study found that Guadua tissue is composed of 40% fibre, 51% parenchyma and 9% conductive tissue.
- Molecular analysis using amplified fragment length polymorphisms (AFLPs) for *G. angustifolia* in Colombia, with emphasis on the coffee-growing area. This study indicated that the natural variability and the genetic diversity of this species in the Colombian coffee-growing region is not significant and that the differences found correspond mainly to the quality of the site.
- Selection, genotypification and multiplication of nine superior materials of *G. angustifolia* for agroindustrial purposes in the Colombian Coffee Region. This research evaluated genetic diversity in the species, described population structure in several Guadua forests, identified and established a genotyping system for nine superior clones,

surveyed the genetic diversity and population structure of Guadua forests and proposed a model for the selection of superior clones of *G. angustifolia* for the purposes of construction, furniture, handicrafts and pre-industrialization.

- Validation of *G. angustifolia* as a structural element for design by the method of admissible efforts. This study sought to incentivize the use of quality bamboo poles, giving engineers tools for design and demanding from the builder quality protocol control of raw material.
- Compression strength parallel to the fibre in *G. angustifolia*. This study indicated that the compression strength parallel to the fibre is 56 MPa in this bamboo and that the average elastic modulus was 17859 MPa.
- Torrefied bamboo pellets for sustainable biomass import from Colombia. This study indicated that Guadua bamboo has potential as a sustainable source of energy generating 20 MJ/kg, and also as a solid fuel. It behaves similarly to other traditional sources of biomass; the high alkali content is unfavourable, but can be solved by additional pretreatment.
- Flora and fauna associated with Guadua forest ('guaduales'). This study reported 115 associated birds and 116 associated plants in the guaduales from the Colombian Coffee region.
- *Catalogue of Colombian Plants: Poaceae: Bambusoideae*. This catalogue listed 114 Colombian bamboo species.

The CBS has developed a strong network for Guadua bamboo formed both of its members and with other Latin American organizations. The society provides a permanent news information source on Guadua bamboo events, products and research through communication online, and has signed alliances with other bamboo international societies, in countries such as Peru, Ecuador and Mexico.

Internationally, the CBS has participated in international workshops and in the World Bamboo Congress. Inside the country, it has organized and provided support for several workshops and national and international seminars on Guadua architecture, engineering, botany, silviculture and standardization. Over more than 10 years the CBS has promoted the International training workshop on Bamboo Industry Sustainable Development organized by MOST, INBAR and the Lin'an Modern Forestry Science and Technology Service Center, and more than 20 Colombian participants have attended this workshop.

Since 2003 the CBS has chaired Technical Committee 178 for the Standardization of Bamboo-Guadua led by ICONTEC (the Colombian Institute of Technical Standards and Certification), and supported the processes of standardization, the key technical tools and clear language needed for improving the quality of processes and products. So far, 14 standards have been conscripted. Since 2015, Technical Committee 178 has been a participant member of the ISO/TC (International Organization for Standardization Technical Committee) 165/WG 12 'Structural use of bamboo' and ISO/TC 196 'Bamboo and rattan'.

The work of associate experts of the CBS in the field of construction with bamboo is very broad and varied. The experience of all of them has contributed to knowledge on the behaviour of the Guadua bamboo as a structural element, and its validation and standardization. In 2009, the CBS, led by Simón Vélez, elaborated the Colombian structural construction code for Guadua and the permit for the inclusion of Chapter G12 in the Colombia code for seismic resistant construction – NSR-2010. This is an important issue because since this code was issued, it has been possible to build houses with one or two floors in urban areas in Colombia using Guadua as a structural element. This is the first Guadua construction code in Latin America and it has served as a model to other countries, such as Peru.

The CBS is a permanent member of the National Consultation Council of the Guadua Chain, the Sectorial Table for Guadua and the Colombian Guadua Federation, contributing in this manner to integrating the actors involved with this resource and towards the comprehensive development of Guadua in the region and country.

With the objective of publishing papers that contribute to spreading and increasing knowledge about bamboos, the CBS has created the newsletter '*El Renuevo*', the web pages www.sociedadcolombianadelbambu.org and www.bambuguadua.org, and published the book *Exotic Bamboos in Colombia*, as well as a manual about how to select superior clones. In addition, the CBS has invited international and national bamboo experts to share knowledge with the members of the society. These have included Prof. Zhu Zhaohua from China, Jorge Morán from Ecuador, Lynn Clark from the USA, Josefina Takahashi from Peru and several Colombian experts, including Simón Vélez, Eduardo Trujillo, Lina Osorio, David Trujillo and Luis F. Lopez.

In 2007, the CBS was operator and founder of the unit of forest management called 'La Esmeralda Guadua Forestry Nucleus – NGLE', formed by nine farms and having a total area of 11 ha in Guadua bamboo. The main objectives for founding the Nucleus were:

- to integrate the area planted with Guadua that had different owners under one umbrella;
- to order and improve the forestry quality of the plantations;
- to ensure a better quality and better price for the raw material;
- to guarantee protection of the water resources and biodiversity;
- to generate employment;

- to provide support and incentives for the research processes; and
- to strengthen the social and entrepreneurial organizations existing in the region.

With this organizational scheme, the Nucleus has assured the sustainability of the resource, carried out eco-friendly management of the environment and achieved greater profitability for the owners.

The CBS is convinced that a better planet is possible by promoting bamboo, and for this, our Vice President was given a prize by Prince Clauss on behalf of the Dutch Government.

3.6.3 Develop a stakeholders' cooperation model for achieving a multi-win scheme

Harmonious partnership among stakeholders

One of the basic conditions for the sustainability of an enterprise is the stable, sufficient and suitable supply of its raw materials, while for the

Fig. 3.96. A bamboo bridge in Colombia (Zhu Zhaohua).

Fig. 3.97. A highway toll station made of bamboo in Colombia (Zhu Zhaohua).

bamboo farmers, a priority condition is a steady and profitable market. The question is how to establish a win–win partnership between the processors and the producers, rather than maintain a simple buyer and supplier relationship? This is an important issue affecting the sustainability of the bamboo sector, and many bamboo producing countries have started to pay attention to facilitating harmonious relationships among stakeholders. In China, there are various partnerships; one type of cooperation is the shareholder cooperatives formed by professional households, and this cooperation among farmers themselves was able to cope with the deficiency in bamboo forest management caused by lack of or ageing of rural labourers and small-scale management. Household members hold joint shares with their bamboo forests, and these forests are then managed by experienced, skilful and interested households. Through this cooperation, bamboo forest management could be scaled up and, at the same time, efficiency was increased, and more labour could be freed to join other businesses. Another type of cooperation is a partnership between companies and farmer households. Next, we provide case studies of Zheijiang Lin'an (Hangzhou)

Kangxin Food Co., Ltd. and Sichuan Qingshen Yun Hua Bamboo Culture Travel Ltd.

Case study 6: Kangxin Food Co., Ltd. – Company + semi-product processing factories + farmer households + raw production bases

Kangxin was founded in January 1999 as a joint venture of China and Japan, and its main products were deep processed boiled bamboo shoots. In 2005, Kangxin's processing capacity reached 23,000 tons, 90% of its products were exported to Japan, South Asia, Europe and America, and the annual sales exceeded 200 million CNY. In 2005, the company had 158 long-term employees and the product sales reached 100%, so the enterprise was very successful. The company had pioneered in Lin'an in establishing the cooperation model of 'company + semi-product processing factories + farmer households + raw production bases'. The components of the model are described below.

COMPANY + SEMI-PRODUCT PROCESSING FACTORY.   Kangxin   contracted   with   48   bamboo   shoot

Fig. 3.98. A bamboo shoot processing factory (Zhu Zhaohua).

semi-processing factories in Zhejiang and Fujian. The company provided standard boiled shoot cans, technical guidance, product codes and quality supervision to these factories and guaranteed the purchasing of the quality semi-products. Through this cooperation, Kangxin could have a suitable and sufficient raw material supply and, at the same time, reduced costs in factory construction and management. The factories can subcontract with farmer households to provide quality fresh shoots. This set-up also ensures the annual incomes of the farmer households in the area. Each primary processing factory can provide 200–300 seasonal employment opportunities each year, in total about 9600–14,000 employment opportunities each year.

COMPANY + FARMER HOUSEHOLDS. This model helps to ensure the raw material supply of the company and, at the same time, protects the benefits of the households from the risks of market fluctuation. Since 2003, Kangxin has established partnership with over 10,000 households in 39 villages and seven towns in Lin'an. Kangxin has also invited experts from the Lin'an Bamboo Industry Society and Zhejiang Agriculture and Forestry University to provide technical consultancies and training courses to the contracted households. With the technical support of experts and technicians, Kangxin was able to establish its own technical demonstration site, where the company applied forest food quality standard control, and introduced the most advanced bamboo forest cultivation and management technologies. The application of these technologies also increased the productivity of the bamboo forests. According to the purchasing contracts signed with the households, Kangxin guarantees to purchase quality shoots from the farmers according to the market price at the purchasing time. If the market price is lower than the normal prices, the company will purchase at protective prices, which would be higher than the abnormal market prices. To date, the total contracted bamboo forest area of the company has reached 7333 ha.

COMPANY + RAW PRODUCTION BASES. Besides the purchasing contracts, Kangxin has also signed another type of contract with farmer households who were not able to or not willing to manage their own bamboo forests. These contracts allowed the valuable transfer of the management rights of the forests to Kangxin. Kangxin then pays a basic rental fee to the farmer household which is based on the average management income of the year, for example, an average annual income of 3750 CNY/ha.

DEVELOP DOMESTIC AND OVERSEAS MARKETS. Kangxin has established sales branches in Kobe and Ibaraki in Japan, and in Beijing, Tianjin and Hangzhou in China. The sales rate is 100% every year.

The overall Kangxin management model outlined above merged all of the stakeholders – the company (including the Japanese investors), farmers and semi-processing factories – into a common interest community. In this community, each stakeholder is independent, but at the same time, they rely on each other's partnership. They are interdependent and mutually beneficial to one another.

Case study 7: Sichuan Qingshen Yun Hua Bamboo Culture Travel Ltd. – promoting traditional bamboo products and culture, achieving prosperity in the local rural area

INTRODUCTION. Yun Hua was founded by Chen Yunhua, who is one of China's National Level Masters of Arts on Bamboo Weaving Techniques. Chen Yunhua's aim was to promote the long-established traditional culture of bamboo weaving in China, and to educate bamboo weaving art personnel. He not only trained a group of Masters of Art, but also trained a large number of poor rural households, which provided them with skills and employment opportunities, and helped them to increase their incomes and gain wealth.

In 1984, Chen Yunhua started a continuing education school for local youths to learn bamboo weaving techniques and related cultural activities, with the support of local government. The purpose of the school was to increase local rural income and promote bamboo art and culture. The beginner course was 3 months long and the master's course lasted over a year. Besides the government subsidies, the school uses the income from sales of the students' products

Fig. 3.99. Participants in a bamboo weaving training course for Tibetans (Chen Yunhua).

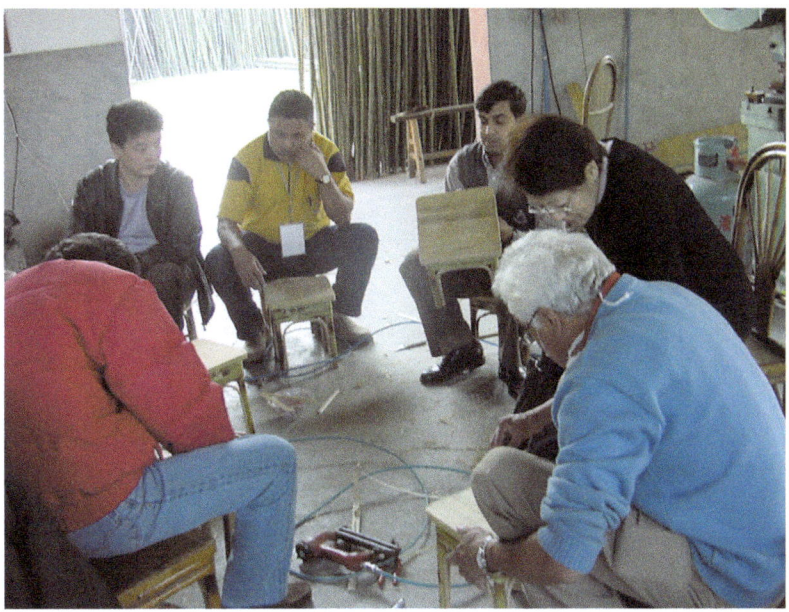

Fig. 3.100. Participants in an international training course on bamboo furniture (Zhu Zhaohua).

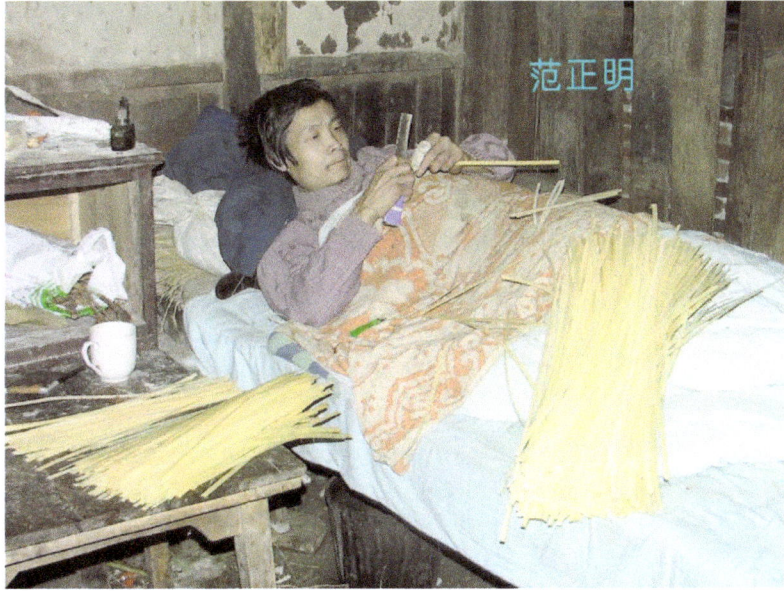

Fig. 3.101. Bamboo weaving by the disabled (Chen Yunhua).

to cover its costs. For each student, the courses and logistics in the school are completely free. Through an effort of more than 30 years, the school has extended its education to other provinces, such as Guizhou, Yunnan, Hainan, Tibet and Guangxi, etc. Over 20,000 young trainees have participated in 120 courses; 80% of these young people came from Sichuan, some of them were from ethnic groups located in remote and poor areas, and others were disadvantaged (disabled)

Fig. 3.102. A fine bamboo woven product (Chen Yunhua).

people. These people were able to acquire a skill and employment opportunity after they had undertaken the courses offered by the school and so were given a chance to get out of poverty.

Chen Yunhua was also entrusted by the Mekong Bamboo Project (sponsored and initiated by the PI and SNV) and the International Centre for Bamboo and Rattan (ICBR) of China's SFA to carry out a number of technical exchange activities and training courses in other countries. The footsteps of Chen Yunhua Masters of Arts have covered Indonesia,

Vietnam, Laos and Cambodia in Asia, Ethiopia and Rwanda in Africa, and Ecuador in Latin America.

As a result of the above continuing education and training activities, Chen Yunhua has trained a number of Masters of Arts, and been able to produce a series of fine art products. As mentioned in Chapter 2, near the end of Section 2.2.4, Chen Yunhua himself had worked with six of his apprentice Masters, taking a whole year, to reproduce the famous Chinese Song Dynasty painting – *Along the River During*

Qingming Festival (*Qing Ming Shang He Tu*) completely by weaving. The whole piece weighs only 300 g but it was sold at the price of 1.06 million CNY. Another piece of work – *A Hundred Emperors in China's History*, was a product of Chen Yunhua and one of his 'disciples', also a Master of Art – Ms Zhang Baozhen, and was sold at the price of US$48,000 in 1991.

In the past 30 years, Chen Yunhua has produced large numbers of exquisite woven products, reproducing China's ancient paintings, famous poems and calligraphy, portraits of famous people, etc. These products have been greatly welcomed by customers in the markets of high-level gifts and art collections. The exquisite products have also promoted China's traditional bamboo weaving techniques and bamboo culture and, at the same time, cultivated a number of heritage persons. In 2008, Qingshen County's bamboo weaving art was recognized by China's State Council as a 'National Level Intangible Cultural Heritage'. Chen Yunhua was recognized as a Registered Master of Crafts and Art by the International Occupation Planning Certification Center (OPC) in 2013, and in 2014, he was granted the title of 'First Class Master of Art' by the China Association of Crafts and Art.

On the basis of traditional culture heritage, Chen Yunhua has also explored innovative creations in the art of bamboo weaving. As a result of his efforts, he has now obtained 22 patents for his innovations in two-dimensional weaving, three-dimensional weaving, fine bamboo weaving over porcelain, woven packages, furniture, interior decoration, bamboo architecture, bamboo chip paintings, bamboo woven purses, etc.

ESTABLISHING A STABLE PARTNERSHIP BETWEEN THE COMPANY AND THE RURAL HOUSEHOLDS. Among the many trainees of Chen Yunhua, those very excellent trainees were directly employed by the company after training. They undertook the production of fine art products or new product development. Most of the trainees returned home, and while continuing to work in the farmlands, they used their spare time to do bamboo weaving and manage microbusinesses. At the same time, Chen Yunhua kept a close cooperative relationship with these former trainees, and when the company has an order, it will talk with the representatives of each community about cooperation in completing the order. The issues

discussed will include sample design, the total number of products ordered, quality control, deadlines and prices. Once agreement has been reached, they will sign a cooperation contract. The professional households would be responsible for the supply of raw materials and weaving, and the company would be responsible for finalizing the products (e.g. lacquering, dying, preservation, etc.), packaging and sales. Every year, the company has brought great economic benefits to the professional households as a result of their cooperation. When talking about their cooperation with Chen Yunhua, the farmers have a very lively description: 'Yunhua has helped us to weave out three-floored villa houses, cars, tractors and TVs.' In 2013, the production value brought by Chen Yunhua to these local rural communities reached 173 million CNY.

SCALING UP AND COMMERCIALIZATION OF BAMBOO CRAFT PRODUCTS. How should the industrialization, scaling up and commercialization of a handicraft product be realized? People usually think it is hard, but Chen Yunhua has made it possible through the partnership model of company + farmer households. Brief accounts of two interesting cases follow.

In 2001, an American customer estimated that the summer in the USA that year would be very hot, and ordered 600,000 bamboo weaving hats from Chen Yunhua. The order was placed on 16 February 2001, and the deadline was in the summer of the same year. For one company producing handmade products, this large order would usually be very hard to complete, but Chen Yunhua did it, and from 16 February to 30 August, the monthly production of the hats was on average 100,000. All of the hats were supplied in time and in good quality.

Over the last several years, there has suddenly emerged a huge market for bamboo woven packages for pine mushrooms, the annual market demand being nearly 100,000 pieces. This huge task was also completed by Chen Yunhua through the cooperation of the company with many professional households. More than 2000 local households worked with Chen Yunhua on this task.

BAMBOO WEAVING PRODUCTION + CULTURE + TOURISM + SALES – DEVELOPING CULTURAL TOURISM. Besides managing the business of bamboo woven products, Chen Yunhua had been developing the

Fig. 3.103. Providing jobs for the disabled working with bamboo (Chen Yunhua).

Fig. 3.104. A production base for bamboo products in a rural area (Chen Yunhua).

Fig. 3.105. Handbags made from woven bamboo (Chen Yunhua).

Fig. 3.106. The Chinese Museum of Bamboo Weaving Art in Meishan City, Qingshen, Sichuan Province (Chen Yunhua).

Fig. 3.107. Bamboo woven products in Guangning, Guangdong Province, China (Zhu Zhaohua).

related cultural tourism in Meishan City, Qingshen, Sichuan Province. The company constructed the 'China Bamboo Weaving Art City', the 'Bamboo Woven Product Museum' and the 'Bamboo Carving Art Corridor', which presented China's historic littérateurs. Through these innovations, Chen Yunhua was able to access the AAA level tourism site requirements and managed a tourism business. Various cultural activities related to bamboo were put up in the China Bamboo Art City, such as bamboo woven products exhibition, a demonstration of bamboo weaving skills by Masters of Arts, a calligraphy and painting studio, a games and DIY room, cultural performances, etc. The Museum was able to collect thousands of traditional and modern bamboo woven products, and showed the long-established history of bamboo culture. The bamboo carving corridor recorded some 448 littérateurs and more than 400 works from them in the period from 770 BC to AD 1911; each bamboo plaque was 1.3 m wide and 1 m high. The annual number of tourists received by Chen

Yunhua's tourism site has reached 200,000. The combination of bamboo culture, with the demonstration of bamboo weaving arts and skills, and sales of the products has widened the market channels of the company.

3.6.4 About expert consultations

When starting a bamboo sector in a country or region which has no or limited practical experience in the sector, inviting experts to provide consultancies could be a good option to quickly build capacity and generate multiple outputs. Based on past experience, countries or regions participating in workshops in China, or Chinese experts visiting those countries and regions to provide training and consultancies, can significantly raise people's awareness and provide constructive support for the identification of bamboo development strategies at national or regional levels. However, these activities may have limited roles in facilitating the detailed practices of a bamboo development project, or the learning of a specific technology. A workshop may not be able to take care of all of the specific needs of the participants. In this case, inviting an expert to work at the project site for a comparatively long period, or sending a technician to study in China or in other countries with mature technologies could be a good option. When inviting experts to provide consultancy services, the following factors need to be considered: (i) the worldwide location of the relevant experts needs to be known; and (ii) the expertise of the experts invited must be closely related to the problems that need to be dealt with.

In the above sections, we have mentioned that representatives from many countries have come to China to study China's experiences, but this does not mean that China is successful in all aspects of the bamboo sector, and has all of the necessary expertise. There are many excellent experts in other countries and regions who can tackle specific technical issues. For example, in Colombia in Latin America, there are many experts on bamboo architecture, especially bamboo pole structures and buildings, and in India, there are mature technologies in the field of bamboo corrugated board.

There is no universal expert that deals with all problems. In the initiating period of a bamboo industry, development strategies and planning are usually necessary, so a bamboo taxonomist, a bamboo enterprise management expert and a development strategy expert are required. These three types of expert could form a consultancy group, and work with local scientists, experts and government officials, to identify a specific strategy according to the particular local conditions. My consultancy experiences in Ecuador, Vietnam, Laos, Bangladesh, Ghana, the Philippines, Timor-Leste and Nigeria have indicated that the combination of these three types of experts can be ideal and effective. Usually, such an expert group needs 15 to 20 days to work out a primary suggestion for a middle/small-sized country on strategic planning for its bamboo sector; this includes an evaluation of the existing bamboo resources, the suggested product structure, the processing technology, investment scale, demonstration site construction and relevant policies, etc.

If there are detailed technical issues to deal with, for example, the recipe for glue composition,

designing a production line, the installation and trialling of equipment and cultivation of plantations, specialized experts need to be invited. Special note needs to be taken that the expert has strong practical experience and operational capacity. For example, if there is a technical problem in processing, it would be very advisable that an expert be invited from a successful enterprise that is working on the same business(es). Such experts have had long periods of first-hand experience, and they will figure out the problem and resolve them in good time. Below are stories of the consultancy of a Chinese expert group for a Vietnamese company, and of a 22-year-old engineer from Anji providing technical consultancy to Ecuador.

Case study 8: report on the consultancy for a bamboo company in Vietnam

Invited by one of the bamboo companies in Vietnam (designated here 'The Bamboo Company'), a team of experts from China visited the company. The expert team was composed of Prof. Zhu Zhaohua and a bamboo engineer, an economy expert and a bamboo processing expert. The team first visited the primary processing (preprocessing) factory of the company, and the manager of the company introduced the production process and stated the major technical problems. The team also visited a preprocessing factory and a local bamboo chopstick processing factory in Lanshan, Thanh Hoa Province, and a plantation of *D. barbatu*. Last, the expert team visited the main processing factory of The Bamboo Company – a composite board factory producing three-layer composite flooring – and had a detailed discussion with the local management staff and technicians on each processing step.

After visiting the primary processing factory and the main processing factory, the expert team provided a detailed report (Zhu Zhaohua, 2006) and made some suggestions for improving the company's production process. The report on the consultancy is provided in Annex 2.

Case study 9: report of Wu Qing's consultancy in Ecuador

Wu Qing, a bamboo machinery expert from Anji, was invited by INBAR-LAC (INBAR's Latin America and the Caribbean Office) to visit Ecuador from 18 July to 7 August 2005. His mission

Fig. 3.108. Bamboo raw material transported by water in Vietnam (Zhu Zhaohua).

Fig. 3.109. Handicrafts made from bamboo slivers in Vietnam (Zhu Zhaohua).

Fig. 3.110. The use of (bagged) bamboo sawdust to grow mushrooms in Vietnam (Zhu Zhaohua).

Fig. 3.111. The Chinese bamboo machinery expert Wu Qing (second from left) working in Ecuador (Zhu Zhaohua).

during the visit was to identify and analyse the bamboo processing machines installed under the aegis of the Ecuabambú project for the conservation and (sustainable) exploitation of Guadua forest in Ecuador and APROGUADUA (Asociación de Producción, Procesamiento y Elaboración de Productos de Guadúa – Association for the Production, Processing and Elaboration of Products of Guadua of Ecuador) in the Bamboo Collection and Processing Center in the village of Carlos Julio Arosemena in the Daule Peripa area of Ecuador. Wu Qing was to make sure that the machines in the processing centre were installed and working properly, and to train a certain number of local workers on the operation of the machines. Below is a report of his work in Ecuador and the suggestions that he made (Zhu Zhaihua, 2005b).

INSTALLATION OF THE MACHINES. According to the conditions of the newly established factory, and considering the convenience of inputting and outputting during processing, Wu Qing instructed the local workers on installing the machines in the proper positions. The machines involved were a raw cutting machine, a splitting (punching) machine, a flattening and width-fixing machine, a primary planing machine, a slicing machine, a fine planing machine, a weaving machine, a blade grinding machine and a coiling machine; as well as these, Wu Qing helped the factory to build a bamboo floor chip slicing machine.

PROBLEMS WITH THE MACHINES. Except for the floor chip slicing machine, the other machines were all imported from China, so had been built mainly according to the properties of Moso bamboo. However, the bamboos in Ecuador are tropical species, such as *G. angustifolia* and *D. asper*, and the culms of these species, in comparison with those of Moso bamboo, have thicker walls and harder, bigger nodes. In this case, if the Chinese manufactured machines were not adjusted accordingly, it would be very difficult to use them for processing. The major problems found are described below.

Rotation speed of the blades. In China, 5600 rotations/min would be fine for Moso bamboo processing, while for the above two local species in Ecuador, the rotation speed of blades should at least be 7000 rotations/min. For a fine planing machine, the rotation speed should reach 15,000 rotations/min.

The edge pressing wheels in the planers. While feeding the bamboo laths into the primary and fine planing machines, there are several sets of wheels that should press the edges of the laths flat. However, it was found that the laths of the local species were so thick that they got stuck at the wheels while they were being fed in.

The springs of the planers. The ranges of the springs in the primary and fine planing machines were not appropriate for the thickness of the local bamboo species.

The matching of machines in the production line. Unreasonable matching of the machines in a production line will affect production efficiency. One flattening and width-fixing machine should be working with two or three slicing machines for best efficiency, but there was only one slicing machine in the factory. Also, the slicing machine should be installed with four or five blades to guarantee a better quality product, and there were only three blades installed.

The weaving machine. The weaving machine could only work with 3.6 mm flat slices or 2.5 mm round slices. However, the blades in the slicing machine were only for 4.0 mm round slices. If the factory wishes to work with 4.0 mm round slices, the material conveying part of the weaving machine needs to be changed, while if the factory wishes to produce 3.6mm flat slices or 2.5 mm round slices, the blades of the slicing machines need to be changed.

Flooring laths. The factory also wishes to produce bamboo flooring laths, and this would need a flooring lath punching machine.

SOLUTIONS.
Rotation speed of the blade axis. The rotation speed of the blade axis is subject to the diameter of the belt. Wu Qing calculated the proper diameter of the belt for the rotation speed required, and found that by changing the belt, the product quality would be improved greatly.

Edge pressing in the planers. In place of the pressing wheels, Wu Qing used pressing plates, while at the feeding end, the plate was shaped to meet the size of the materials. Feeding in the laths was then easier.

Springs. It was decided that longer and stronger springs would be purchased from China to meet the needs of the local species.

Matching of machines in the production line. A four-blade slicing machine should be added to the production line. The present three-blade slicing machine could only work on 3.6 mm flat slices or 2.5 mm round slices, and if a four-blade slicing machine was added to the production line, the factory would be able to produce various sizes of round-shaped slices for use in producing curtains, toothpicks and skewers.

Flooring laths. A bamboo flooring lath punching machine was built under Wu Qing's instructions, and is now working in the factory. However, the cutting saw of the machine purchased from local market was too thick – 3.0 mm. If a 1.5 mm saw was installed, the wastage rate of raw material would be reduced from 16 to 9.6%. Because a 1.5 mm saw was not found in the local market, it is suggested that the factory import this saw from China.

Moisture content of the materials. G. angustifolia has a higher moisture content than D. asper and B. vulgaris, and it was extremely difficult to feed this material into the machines. The slices were crude and crispy. Following several tests, this problem was solved: after splitting, the fresh bamboo materials were not processed directly, but left to dry naturally for 2 days; after this, they were ready to be processed into quality slices or flooring laths. In contrast, D. asper and B. vulgaris could be processed freshly. After further testing, it was found that B. vulgaris was the best material for the processing of bamboo slices.

TRAINING LOCAL WORKERS.   The 3 week demonstration and training by Wu Qing achieved satisfactory results.

Operation of the machines. A technician was trained on the maintenance of the machines.

Each machine had at least one worker, while each slicing machine had two workers. All workers had mastered the operating skills of his/her machine, and certain debugging methods. However, the training period was not enough for the workers to become proficient in all the skills taught by Wu Qing. According to him, usually it takes 3 months for a new worker to become skilful in the operation and simple debugging of these machines.

Training on the processing

1. Cross cutting: grading the materials according to diameters could increase the efficiency of working and reduce wastage of materials.
2. Splitting on the punching machine: this required the selection of different blades according to the diameter of the cuttings (tubes).
3. Flattening and width-fixing machine: grading the laths before feeding them into the machine would help to improve work efficiency.
4. Slicing: the better the thickness of the laths meets that of the slices, the more economic it is for the utilization of material.
5. The slices should be treated with H_2O_2 solution: the proper concentration for the treatment should be 25–30%. The H_2O_2 solution sold in the local market had a concentration of 50%, so it needed to be diluted before use. For example, if the treatment tank is 80 × 80 × 220 cm, add water to 80%, then add 10 kg of H_2O_2 (at a concentration of 50%), heat the solution to 70–80°C and dip the bamboo slices into the solution for 10–15 min (or judge by the naked eye when the slices turn to a white colour, when the treatment is done). Flooring laths need to be treated in an 80–90°C solution for 2 h. A treatment tank was built in the factory as per these suggestions from Wu Qing.
6. Drying: the slices can be dried under the sun for 2 days or in a hot air kiln during the rainy season. A kiln and a drying room were designed by Wu Qing and installed in the factory. The drying room has a volume of 16 m³ (2 × 2 × 4 m).
7. Security: the factory should be a restricted place where children are not allowed to enter. Abnormal touching of the machines was not allowed, and standard operations were required. The operators were not allowed to clean the sawdust and chips on the machines directly with their hands. All of the switches on the machines must be checked before the power was switched on.

Three species normally seen in Ecuador were selected. As already mentioned, *G. angustifolia*, *D. asper* and *B. vulgaris* were selected for testing on the machines. The first two species were suitable for processing flooring laths as well as slices for sticks, curtains, etc.; *B. vulgaris* was not suitable for flooring laths (the walls are too thin), but can be processed into slices for curtains, mats and sticks.

Special adjustment for Guadua. According to the properties of Guadua bamboo materials, Wu Qing suggested that the splitting machine (puncher) should use a 6P engine, which has a rotation speed of 960 rotations/min, but a stronger power than the present engine – a 4P engine with a rotation speed of 1400 rotations/min. Wu Qing also suggested that the engine for the slicing and fine planing machines should be a 4P instead of the present 2P engine. By calculating the diameter of the belt connected with the blade axis, the rotation speed of the blade axis in the slicing machine should reach 5000 rotations/min or above, while the rotation speed of the blade axis on the primary and fine planers should be above 7000 rotations/min, and 10,000 rotations/min would be best. Also, of course, before processing, the fresh splits should be dried naturally for 2 days. The above methods could guarantee that quality Guadua bamboo slices and flooring laths are produced.

Sawing for material preparation. The sawing methods for fresh bamboo materials should be adapted according to the requirements of the products, and the thickness of wall and the length of the bamboo culms, as reported in Table 3.4 for Guadua and Table 3.5 for *D. asper*.

For *B. vulgaris*, as the culm walls are comparatively thinner than those of Guadua and *D. asper*, and the thickness is even from the bottom to the top of the culm, it was suggested that this material could be cut into 1.25 m long pieces, and 5–6 pieces per culm.

Table 3.4. Sawing requirements for *Guadua angustifolia*.

Culm properties	Base part	Middle part	Top part
Thickness of wall	1.3–1.4 cm	0.9 cm	0.5 cm
Percentage of the full length	15–20%	70%	10–15%
Length per piece	1.6 m	1.6 m	1.6 m
Number of pieces per culm	1–2	4	1–2

Table 3.5. Sawing requirements for *Dendrocalamus asper*.

Culm properties	Base part	Middle lower part	Middle upper part	Top part
Thickness of wall	2–3 cm	2 cm	1 cm	0.5 cm
Length per piece	1.6 m	1.6 m	1.6 m	1.6 m
Number of pieces per culm	1	1	12	2

The suggestions made above will improve the utilization rate of the three species of bamboo materials.

MORE SUGGESTIONS

About the machinery

1. The factory needs one more manual bamboo splitter, and the local workers could make this themselves.
2. The factory needs one more slicing machine which has four blades.
3. In order to produce bamboo skewers or toothpicks, the factory needs a sharpening machine, a grinding machine and a slice cutting machine (these can all be self-made).

The main products. The available local species and existing machines can produce bamboo laths for flooring and laminated furniture, slices for curtains and mats, and sticks for skewers and toothpicks.

Training on bamboo machinery. It is suggested that a Chinese machinery technician be sent to Ecuador to provide training for local technicians and workers. The duration of the training should be no less than 3 months and no longer than a year. Conversely, Ecuador could send technicians and workers to China for the training.

Importing bamboo machines and equipment from China. All of the machines and equipment should be adjusted according to the properties of the local bamboo species before shipping from China, in order to make sure that the machines work properly after installation in the factories in Ecuador. Because the voltage in China is different from that in Ecuador, the motors can be accessed locally or configured before shipping.

Discussion on case studies 8 and 9

ABOUT THE CONSULTANCY OF CHINESE EXPERTS FOR THE VIETNAMESE COMPANY: CASE STUDY 8

1. The consultancy is targeting a newly established large-sized company. The expert group is composed of a machinery engineer, a bamboo material expert, an economist and a bamboo industry development strategy expert. Although the consultancy period was rather short, the expert group provided a deep, detailed and well-targeted analysis of the problems of the company and suggested solutions. This was the advantage of a comprehensive expert group.
2. The evaluation report of the expert was an all-in-one document. It did not only point out the existing technical problems, but also provided solutions for the company's production and business management, sustainable raw material supply and reform of the salary scales for the workers.
3. Because of the serious management problems of the company, the suggestions of the experts may bring a complete change to the whole company. According to a preliminary estimation, if all the suggestions made are adopted, the company's production efficiency will be increased by 60% after a year, while the product costs will be reduced by 40–60%. The raw material utilization rate will be increased from 6–8% to 50%. The company would then be able to sustain itself.
4. However, for various reasons, the company did not adopt the experts' suggestions. This was a pity.

THE CONSULTANCY OF WU QING: CASE STUDY 9

1. Before the equipment was imported, the Ecuadorian side had not actually made a clear study of the properties of the local species and of other natural conditions that might affect the production. Apart from this, there was little knowledge on the properties and functions of the imported equipment, even the power voltage that was needed, and when the machines arrived, it was very hard to enter into real production.

2. Before importing the equipment, there were no tests on the properties of the local bamboo, for example, the thickness of the culm walls, the moisture content of the fresh materials, the hardness of the node parts, etc. The equipment was not adapted for the special properties of the local bamboo species, which made it hard to work it. Such problems were hard and it took a long period of time for the local technicians of Ecuador to explore solutions.

3. Wu Qing had only 20 days for his consultancy, but he solved all of the major technical problems. Nevertheless, it was estimated that more problems would appear after his consultancy, so the last two points that he made – on (further) training on bamboo machinery and importing machines and equipment from China – were very important.

Packaged consultancy service

The case of Vietnam made me think about the many types of consultancy services that could help a newly established enterprise to acquire management and production capacities quickly, and a packaged service could be one of these. Based on the needs of a company, the packaged service could arrange to provide production management, product orientation, production line design, equipment selection, installation and maintenance, personnel training for managers and operation workers, etc. The packaged service could be provided by an expert team composed of a production manager, a technician (engineer) and a bamboo industry development expert. A short-term contract (e.g. 1 year) could be signed with the team to provide consultancies on all of the aspects listed above, and when the whole company's production and management was going steadily, the consultant group could leave. Could this method/approach be trialled?

3.7 Market Development

Enterprises conducting real production need to consider the potential market for their product(s) and evaluate the market demand. An investigation on the market demand would be a first step in market development. The easiest approach is to start with local and domestic markets. For international market, the enterprises need not only to investigate the potential demand, but also study whether there are potential competitors and research on strategies/tactics to cope with potential risks. The development of unique and featured products that are easily accepted by customers is key to winning markets. Control of the logistical costs and shortening of the chain of sales are keys to increasing product competitiveness. Therefore, a healthy, growing and freely competitive market environment is very important for the development of an industry.

Over the last 30 years, China's bamboo sector has been developing steadily, continuously, fast and sustainably. One of the reasons for this is that the related government departments made market construction a prioritized mission for sector development; another reason is the continuous efforts of the enterprises in optimizing product structure, innovating new products and opening up new markets.

3.7.1 Market development of the bamboo sector in China

Before the reform and opening up-policy, China's bamboo sector, except for the pulp industry, basically involved traditional manual production, with low value and low productivity. The main products were bamboo poles for construction (scaffolding), transportation (bamboo rafts), daily used handicrafts and fine art products. Another category was bamboo shoot products for food, with a small amount used for medicine production. Since 1985, when bamboo processing machines were introduced to Mainland China, the bamboo sector has been developing fast, and the categories and types of products, the production scales and the production value have all increased several fold. Figures 3.112 and 3.113 show that from 1990 to 2012, China's annual bamboo production value increased

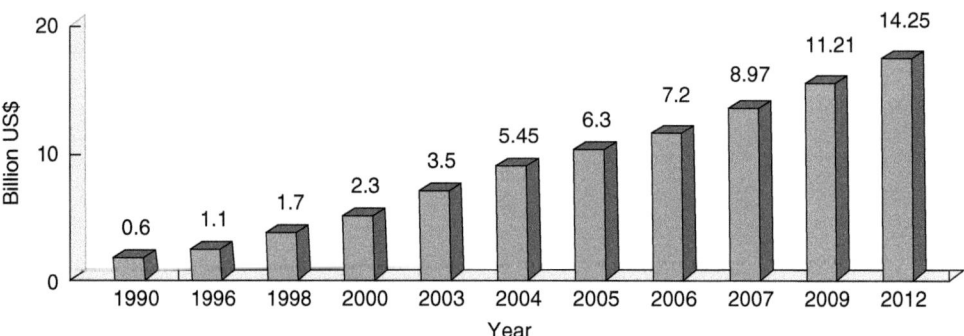

Fig. 3.112. Annual production value of bamboo products in China (1990–2012).

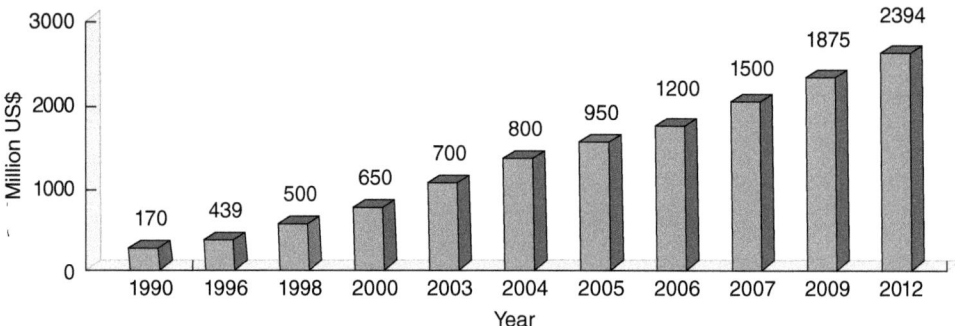

Fig. 3.113. Annual export value of bamboo products in China (1990–2012).

23.75 times, and the annual export value increased 14.08 times.

Case study 1: the domestic market – Anji's market development experiences

China is a huge country with large population and long history of bamboo cultivation and utilization, and the people have a natural sense of closeness to bamboo, so the domestic market for bamboo is huge. However, to deliver bamboo product smoothly to customers, the market needs to experience a period of gradual development. Here I would like to tell the story of Anji.

In the initial stage of bamboo development, some bamboo farmers grew into middlemen who managed a microbusiness or a small-sized business. They purchased bamboo products from local factories at wholesale prices, and set up sales shops in different parts of the country. Merchants from other locations also came to the producers to purchase products. At this stage, the middlemen were very effective in sales, especially for products used daily such as chopsticks, curtains,

mats, handmade daily crafts, processed shoots, etc. This type of sales could be called the 'ants moving' model. Through 'ants moving' activities, the products got sold, the middlemen grew and increased their sales capacity and, at the same time, earned money. The middlemen were not only conducting sales, they were playing the roles of information exchange. In turn, their intime feedback to the factories helped the production enterprises to improve their products.

LOCAL PROFESSIONAL MARKETS AND TRADE CENTRES. In the 1990s, several bamboo product sales markets were established as per the requirements of processing enterprises and under the support of the local government. At the beginning, two streets in the downtown area of Anji were identified as bamboo trade streets. The production enterprises could rent the shops along the streets for product sales, and the shops conducted wholesale and retail selling at the same time. The development of the bamboo trade streets made it easier for merchants from other locations of China, and even for merchants from overseas.

In order to meet the new requirements of the fast-developing bamboo sector, in 2007, a large-scale bamboo product trade platform was established as a result of public–private partnership. The China (Anji) International Bamboo Trade Mall took a total area of 35.1 ha with more than 1200 shops. Since its opening, the trade value of the mall has been increasing year by year. It not only provided a market platform for local enterprises, but also provided services for enterprises from other bamboo producing areas of China. The trade value of only one type of product – bamboo curtains and mats (mats, curtains, carpets, table mats, pillow mats, car seat mats, etc.) reached 6.3 billion CNY in 2013.

TRADE FAIRS AND EXPOSITIONS. Besides the above-mentioned two sales models – the 'ants moving' and 'trade centre' models – there have been many other types of bamboo product trade platforms in China, and enterprises were free to choose to participate in any of them. Examples include the national furniture exposition, international construction material trade fair, the national craft and arts exposition and food trade fairs, etc. Some of the platforms were nationwide and were closely related with the bamboo sector. Usually, for these types of fairs and expositions, the related government sectors will intervene and organize the participation of bamboo enterprises. For example, the China National Forest Product Exposition is an annual event sponsored by China's SFA, and the SFA would usually require provincial departments of forestry to organize the participation of enterprises. This exposition has always been a large-scale and comprehensive event that gathers tens of thousands of different products from the forest, such as wood products, non-wood products, foods, medicines, arts, etc. In 2013, Anji Forestry Bureau (2013a) took the lead in organizing local enterprises to participate in the event, and obtained 34 booths. In the same year, three enterprises from Anji were able to put up an exhibition called 'Green Rhyme – Bamboo Music, Culture and Art' in the China Grand Theatre in Beijing.

INTERNET MARKETING.   In a similar manner to the other trade sectors of China, the online trading of bamboo products has been taking on a more and more important role since the development of the Internet in China. In Anji, almost all enterprises producing final products have their own websites. Online sales have become a very important way of marketing for the bamboo enterprises.

Case study 2: the international market –Anji's market development experiences

In the initial stage of bamboo sector development in China, enterprises attached great importance to introducing their products to the international markets. The reason was that China had just started to implement its opening up policy and there was a big demand for foreign currency, so exporting products could help earn foreign currency and also to purchase advanced products from developed countries. Another reason was that the industrial bamboo products, such as bamboo flooring, were seen as new emerging products for which the domestic market had not yet matured, although these products had raised great interest in the international markets. Many enterprises were by then interested in exporting. However, at the beginning of the opening up policy, the entrepreneurs had only just transferred their enterprises from the planned economy to the market economy, and some of them had only just turned from farmers into business owners, so they knew little about international market development. Here, we take the example of Anji again to introduce the international market development of China's bamboo sector.

'HENS AND CHICKS'.   From the late 1980s to the early 1990s, bamboo enterprises lacked experience in and channels for exporting. On the one hand, they were still bound by the concepts of the planned economy, and on the other hand, the enterprises were mostly micro- and small-sized homestead factories. Their foreign trade businesses were managed by the Zhejiang International Trade Co., Ltd., a trading company that is a state-owned enterprise, and is responsible for negotiation and contracting with foreign traders, and helping in foreign trade procedures.

The trading company would sign a product supply contract with the producing enterprise before they signed a contract with a foreign customer, in order to ensure that the products were supplied in time and were of the right quality and quantity. While they were contracting with the trading company, the bamboo enterprises in Anji started to learn about foreign trading procedures, the demands and trends of foreign markets, and the standards needed for products and packages. In this way, they gradually grew their businesses with foreign customers and, at the same time, strengthened their own foreign trading capacity.

As the bamboo sector grew and more products were exported, the state-owned trading company was no longer capable of meeting the demands of the growing businesses, and since the early 1990s, the local government allowed a few enterprises with a big exporting value and higher quality personnel the rights to independently export. Those enterprises without such rights started to entrust the enterprises that had obtained exporting rights to manage their foreign trade businesses.

As the amounts of exports continued to grow, more enterprises grew bigger and accessed their own exporting rights. This process of growing, in which large enterprises lead small enterprises, I would like to call the 'hens and chicks' model; this model had cultivated a group of farmer entrepreneurs who were good at foreign trade and international marketing.

THE NATIONAL FOREIGN TRADE PLATFORM. China has large numbers of foreign trade platforms, and they have provided convenient conditions for enterprises to show and export their products. The one that has the longest history and largest scale is the Canton Fair, which was started in 1957, and is held twice a year, in the spring and autumn. Up to 2013, the Canton Fair had been organized 114 times, and total number of booths had reached 60,222. Every year, a total number of 396,412 exhibitors from 212 countries and regions in the world join the fair, and its annual trading value has reached US$67.23 billion. Many bamboo enterprises met their new customers and established long-term partnerships at the fair. This type of fair can play an important role in the initial years of an enterprise.

INDEPENDENT EXPORTING HAS BECOME A MAJOR MARKETING CHANNEL FOR BAMBOO ENTERPRISES. Having gone through the 'hens and chicks' period, and with experience from trade fairs, bamboo enterprises had become familiar with the regulations and procedures of international trade and, at the same time, they had cultivated a group of specialized personnel and established business partnerships with foreign traders. Most of the bamboo enterprises in Anji started to manage their own exporting businesses and established their own exporting channels. Some enterprises had even established sales offices overseas, or developed their own agencies. Up to the late 1990s and early 2000s, most bamboo enterprises in Anji were exporting through their own channels, and although they still sometimes need to selectively participate in trade fairs and exhibitions in China and overseas, the total frequency of and initiatives for joining such events have been reducing. Instead, a group of entrepreneurs from Anji, who used to be farmers and had only junior level education, were seen to be busy flying back and forth in Europe, North America, Japan, etc., selling their products.

Since around 2005, the situation changed again. The second generation of these farmer entrepreneurs, who had received a higher level education – some even had studied abroad – came back and participated in the businesses. Many of them then inherited the businesses from their parents. These new generation entrepreneurs were advantaged by having modern scientific knowledge, and they were able to master and adapt to the complicated and challenging situations in the international market, and also to master new marketing skills. Their disadvantages were that they lacked the persistence and spirit of working hard that their parents had; they were also reluctant or lacked the capacity to communicate well. Up to 2014, it could be seen that the handover from generation to generation of the bamboo businesses in Anji was partially completed.

3.7.2   Follow and proactively adapt to market changes

An in-time prediction of the changes in the markets could be vital for the survival of an enterprise. This is also a key factor affecting the success and failure of a company.

Case study 3: bamboo panel production development and its changes in Anji and Lin'an

From the end of the 1980s to the early 1990s, bamboo panels, especially bamboo flooring products, developed fast, and the scale of production grew tremendously. Because bamboo materials, compared with (tree) wood materials, were considered more environmental friendly, and also because they were a new type of product at the time, bamboo flooring products won a lot of interest in the international market. Bamboo floorings sold well in the international market, and as the raw material, labour and production costs were comparatively low at the time, the processing enterprises received high profits.

However, as the production scale extended too fast, the raw material and labour costs soared (see Table 3.6), the competition among enterprises intensified and a chaotic low-price competition appeared. At the same time, the CNY exchange rate rose. Also, because of the limited portfolio (a single product – laminated flooring), the interest of international customers started to drop. From Table 3.6, it can be indicated that in 2006–2007, the FOB (Free on Board) price of the standard flooring tile (920 × 184 × 15 mm) was very close to its production costs, the profit rate was only 6.6%. This was a huge drop from a figure of 191% in 1992–1993. By 2011–2012, the enterprises producing and exporting laminated flooring were already losing 10.2%.

In the face of these changes in the international market, and the new situation in the domestic bamboo processing industry, some enterprises had by 2002 already started to adjust their products and production lines to meet the new market needs. They adjusted their production lines from the production of a single laminated product to the production of multiple laminated products. The new lines could produce various types and sizes of laminated panels – the length could be 2-4 m, the width from 18 to 20 cm and the thickness could be adjusted from 0.8 to 2.0 cm. These other types of laminated board were by then called 'special-specification laminated panels'. The biggest advantage of such panels was that they can not only be used for flooring production, but also be processed into decorative boards and furniture boards, so they have a much wider adaptability. When these new products entered the market, they soon raised the attention of customers, the sales soon started to increase, as well as the prices, and they started to win over the laminated flooring markets. Those enterprises that had first adjusted then got the priority in market development opportunities.

Up to 2003 and 2004, some enterprises started not only to produce specifically sized boards, but also added production lines of bamboo veneer (thickness 0.2–0.6 mm). The veneer products could be used for surfacing furniture, or for handicrafts or the interior decoration of cars. This new product further promoted the value of bamboo panels.

At the same time, the once populated but rather complicated laminated panel technology was gradually being replaced by the much higher efficiency technology of pressed bamboo. The pressed bamboo technology was invented by an enterprise in Anji in 2000. Up to 2008–2010, most panel manufacturers in Anji transferred from applying laminate technology to applying pressed bamboo technology. In 2012, the total number of pressed bamboo production

Table 3.6. The average costs and prices of bamboo flooring board (horizontally and vertically pressed) for each year from 1992–1993 to 2011–2012 (data from Zhu Zhaohua and Jin Wei, 2006).

Year	Raw material cost/m² (CNY)	Product cost/m² (CNY)	FOB[a] price (CNY)	Salary of worker (CNY/month)	Exchange rate with US$
1992–1993	52.1	86.04	250	500–600	1:5.5
1995–1998	56.1	92.7	220	800–1000	1:8.7
2001–2003	59.6	98.43	145	1100–1200	1:8.2
2006–2007	62.5	103.2	110	1600–1800	1:7.5
2011–2012	63	123	110.5	2800–3100	1:6.7

[a]Free On Board.

lines in China had reached 150. The strand-woven technology had the following advantages:

1. The bamboo strips can be directly pressed into lumber that is quite similar to (tree) wood lumbers.
2. Bamboo lumber had wider adaptability than wood lumber, and could be processed into almost anything that wood can be made into. Bamboo lumber can not only be processed into floorings, but also into furniture and construction materials.
3. Bamboo lumber has better abrasion resistance, corrosion resistance and pest resistance than wood lumber, as well as better strength, including tensile strength. This is especially favourable for outdoor constructions.

Through a series of innovations, transformations and promotions, pressed bamboo lumber panels have become widely accepted in the market, which was great progress in the market development for bamboo-based panels, as well as from the aspect of value addition.

3.7.3 Suggestions on raw material markets

A sufficient and steady raw material supply to bamboo enterprises is not only a basic condition for the development of the bamboo sector, but also a key factor that sustains the benefits for the bamboo producers. In the initial stage of bamboo sector development, great importance should be attached to the relevant government departments and enterprises. In preceding sections, we have described the construction of bamboo shoot markets in Lin'an. However, the supply and sales of bamboo culms were quite different from that of shoots. This was because the shoots were food products, and the fresh shoots could be directly transported and sold to end markets. In contrast, raw bamboo culms need to be sold to processing enterprises, especially primary processing enterprises. Let us still take Anji as an example, to explain the history of bamboo raw material market development.

The failure mode of supply – enterprises directly purchase from the vast farmer households

In the late 1980s, Anji experienced a fast transformation from traditional manual production

to modern mechanical production of bamboo products, and the consumption of bamboo raw materials soared. However, the purchasing methods were still the same: in the period from 1986 to 1991, the factories went from one household to another to collect culms or the households transported them to the factories. The transportation costs were high, while the utilization rate of the raw materials was rather low, with some 70% of raw materials wasted. Large amounts of waste materials piled up from the factories, and it was hard to dispose of them (see Figures 3.30–3.32 for examples from Vietnam and China). The factories were overwhelmed.

There was also an example later on, during 2003–2008, when an Anji enterprise established a pressed bamboo factory in Yuxi, Dehong Autonomous Prefecture of Yunnan Province. The factory had planned to utilize the rich tropical bamboo resources of *D. giganteus* and *D. brandisii* in the area. The factory was successfully established, but after a period, production was forced to stop because of the insufficient raw material supply. This was because the raw materials had been planted by different households on their own small-scale forest lands. Each household could only supply a small amount of raw material at any one time, and the factory needed to negotiate with each of them every time when they went to collect the bamboo. The raw material supply processes were too complicated and the costs were rather high. Therefore, the enterprise–farmer household mode of supply was a failure.

Middlemen and the raw materials market

Some of the bamboo farmers were conducting the transportation of bamboo raw materials, and they were familiar with bamboo producing households and their bamboo forests. These farmers became the middlemen for the supply of bamboo raw materials. A commonly agreed site was identified as the raw materials trade market, and the middlemen and factories gathered at the site to conduct trade. This method guaranteed a stable supply of raw materials to the factories, and they could also select the best quality raw materials for their purposes while, at the same time, saving costs. The producing households were also able to benefit. They could sell directly to the raw materials markets via the middlemen, and negotiate with them to pick the best price bid. The middlemen + raw materials market mode of supply provided good employment opportunities for many farmers and also met the needs of the enterprises, and it operated well from about 1991 to 2004.

Anji's annual consumption of Moso bamboo at this time was around 130 million culms, and of these, about 50 million were processed in Anji, and the total weight could be 700,000 tons. Thus, the daily average consumption of raw bamboo materials could be 2400 tons. Such a huge demand for supply was guaranteed by the middlemen + raw material market model. On the road to Tianhuangping Town, every morning before 8 a.m., one could expect to see a spectacular scene: hundreds of tractors and trucks fully loaded with bamboo culms parked alongside the road, sometimes lined up along several kilometres; most of them had come from other counties nearby. After 8 a.m., the staff from the primary processing factories came to select their raw materials; after the deal was made, the staff would lead the tractors or trucks to their own factories. Some of the tractor suppliers got a long-term supply contract, and these supplies were directly transported to the factories. At 9 a.m., this scene was empty, and all of the raw materials were sold. This was a very efficient logistical mode.

3.7.4   Case study 4: innovating high quality products, exploring both domestic and international markets – Ningbo Shilin Arts & Crafts Co., Ltd.

Founded in 1978, this private enterprise was started by Wang Jianqin as the Yanzhen mahogany craft factory under the names of Arts & Crafts Co., Ltd. (Shilin's predecessor) and the Redwood Craft Factory (the predecessor of the existing Scholastic, Inc., an R&D arm of the company); the company was also operated by Wang Jianqin. It started with the production of redwood furniture and wooden chopsticks using imported precious mahogany [most probably a species of *Toona*] from Vietnam, Burma (now Myanmar) and Laos as raw materials. After 10 tough years of wood product processing, Wang Jianqin not only found it gradually more difficult to

import raw materials because of the rising prices, but also considered the industry to be contributing to the destruction on the forest, which is not friendly to the environment. Meanwhile, it occurred to him that in Fenghua County, under the municipality of Ningbo City in Zhejiang Province, where his company was situated, there was a famous bamboo town. Here, a large area of bamboo of high quality had not yet been fully utilized, not to mention industrially processed. Further, in contrast to mahogany, bamboo has great self-renewal capacity, and is the kind of resource that is easy to sustainably manage. So why not use the rich local bamboo resources and produce environmentally friendly products from bamboo instead of wood?

In 1994, the first upgrading and restructuring of Wang Jianqin's enterprise took place. Chopsticks in bamboo came to his mind first. During that time, there was only one type of bamboo chopstick on China's domestic market, 'TianZhu' chopsticks, made from small-sized bamboo; these had rounded ends, and the added value of this type of product was quite low. Wang Jianqin chose to develop 'high-end' bamboo chopsticks targeting the Japanese market, as Japan is the largest market in the world for chopsticks besides China. After a thorough study of the chopstick market in Japan and the largest chopstick producer in China, Shilin Arts & Crafts designed and produced high-grade chopsticks in various beautiful forms, with square or rounded heads and sharp ends. Welcomed by the Japanese customers immediately, the products soon took over the market. At that time, the company produced more than 20,000 pairs of chopsticks a day, and the annual output value was 500–600 million CNY.

Nevertheless, under the serious challenges brought about by the Asian financial crisis in 1998, the Japanese clients started to return the goods one after another, and this had a great impact on the company. However, in the face of the serious impact of the international market, the company did not shrink back. At that time, the biggest difficulty Shilin faced was its capital turnover. To open up the domestic market, Wang Jianqin even sold his own car. At the same time, China's domestic market was experiencing a golden time and further opening up, as the domestic markets, large-scale high-end supermarkets and department stores started to be accessible to private enterprises. This provided an opportunity for Shilin to enter the high-end markets in China. After 2 years of endeavour, Shilin opened up more than 800 agencies across the country, and the superior quality and new designs of Shilin's Japanese-styled chopsticks had won over other traditional chopsticks on China's domestic market, and were favoured by consumers. As a result, the annual output value of the company gradually increased from 8 million CNY to 20 million CNY.

Because of their innovative designs, Shilin's chopsticks were not only beautiful but also practical, and other companies soon started to follow its model. Therefore, a revolution was triggered in the Chinese chopsticks industry, and bamboo chopsticks started to replace wooden chopsticks. At present, bamboo chopsticks make up 70% of the country's market share, and have mainstreamed the chopsticks industry from wood based to bamboo based. Thus, there was a large reduction in the consumption of wood timber, thereby protecting the forests.

After the opening up of the domestic market, Shilin encountered chaotic competition in the chopsticks industry. Many of Shilin's new products were copied by peer companies, which intensified the domestic competition. In order to win the fierce competition in both the domestic and the international markets, Shilin did not stay with its original advantageous products – bamboo chopsticks, but vigorously developed other new, pro-market products. Since 2003, Shilin has gradually developed and produced bamboo salad bowls, bamboo tableware, bamboo cutting boards, bamboo bathroom products, bamboo kitchen knife holders, bamboo furniture, etc., so that it now makes a wide variety of bamboo products on the domestic market that are replacing wood products. Take the chopping boards and kitchen knife holders for instance; in the past, 100% of chopping boards and kitchen knife holders were made of wood, but now 80% have been replaced by bamboo products; the bamboo chopping boards and kitchen knife holders are also exported in large quantities. Shilin has continuously played a leading role in the development of a series of bamboo products, and up to 2005, its output value increased to 30 million CNY. While developing its products, Shilin also put great efforts into exploring the domestic and foreign markets, and established partnerships

with some related companies in Europe, the USA, Japan and so on. Furthermore, the company has participated in exhibitions in Germany, the USA and Hong Kong. In the meantime, they also carried out sales cooperation in major shopping malls in major cities to continue exploring the domestic market.

In order to access international high-end markets, Shilin began to negotiate with IKEA in 2008. It took 4 years and cost 20 million CNY for Shilin to meet all of the strict and comprehensive industry standards of IKEA, which include product quality control, enterprise management, the protection of employee rights, resource management, etc. In 2012, Shilin became IKEA's first Chinese bamboo product supply company. The output value of IKEA products in that year was only 2 million CNY, but by 2015, this figure had increased to US$20 million, and the IKEA product demand has been growing rapidly. In the fiscal year from 2015 to 2016, the total output value of Shilin reached 200 million CNY. The company was awarded the designations of 'The China National Leading Enterprise for Bamboo Sector' and 'China Quality Credit Enterprise' by the Chinese government, and plans to achieve a total output value of 300 million CNY in 2018–2019.

Wang Jianqin used one phrase to sum up Shilin's successful experiences: 'Strong capacity internally, great dynamic externally, …'. For the products to be able to occupy the market as they do, the company needs to make all efforts to strengthen its capacities in all aspects and be ready. Continuous efforts in developing and promoting high quality, innovative products, and the adherence to credibility and integrity are essential conditions for success. With these principles, Shilin's products – bamboo laminated panels, furniture, craft products in daily use, bathroom products, tableware and storage facilities – have been developing from small to large, from simple to fine, and have kept a leading position in both domestic and foreign markets. Presently, Shilin has just completed the development of a new product – a Cellular Board or Hollow Panel, as well as the one-time moulding technology for bamboo furniture components. These two technologies will greatly promote the further development of bamboo furniture and bamboo construction. Shilin attaches great importance to the construction of its innovation team: on the one hand, the company cooperates with China's colleges and universities; on the other hand, it has also invited foreign designers from Germany, Italy and Sweden to join the team. What is more important, they are to pay attention to personnel development, provide stable conditions for experienced employees, allow sufficient space for their innovation activities and to develop a number of first-class craftsmen and Masters of Arts in processing techniques and product designing.

In terms of market development, Shilin has begun to explore new models of Internet sales, and has now got the preliminary results; the monthly online sales have reached 200,000 CNY, but the goal is 50 million to 100 million CNY in 2 or 3 years. Shilin is now cooperating with Ningbo Dahongying University, and has established an on-campus studio, which allows the students to participate directly in the online sales of Shilin's products, while at the same time improving their capacity. The company is also learning e-commerce technologies, as well as training and attracting talents.

Shilin, which has always aimed at developing bamboo household products that could replace or win over those made of wood, has been persistent in its efforts to explore domestic and international markets, and continues to present environmentally friendly products to its consumers.

3.8 Pay Attention to Bamboo Product Innovation and the Sector's Life Cycle

3.8.1 The law of products and the industry life cycle

Neither a single product nor a special industry lasts forever. Each product and industry goes through phases of growth, maturity and, finally, decline. For instance, bamboo has a long-standing history of utilization in China that has been growing continuously for 7000 years. People used to use baskets made of bamboo on a daily basis, but as plastic products became more common, they lost interest in learning and preserving handmade ways of artistry. Furthermore, before the 1970s, it was a common site to see

people carrying bamboo baskets, crates and boxes everywhere in the southern parts of China, but by 1990, plastic products had replaced these more traditional forms of bamboo products.

Another example of decline is the period of the years 1986 to 1995 when bamboo stick products (toothpicks, skewers and chopsticks, etc.) were an integral part of Anji's day-to-day life. Later, due to the increasing prices of raw materials and requirement of high labour wages, other products, such as bamboo mats, bamboo curtain, and panels began to appear and the share of bamboo stick products gradually declined. Production centres moved westward towards provinces such as Jiangxi and Hunan, etc. where the price of raw materials and labour is relatively cheap.

After 2000, the production centres of bamboo sticks moved again, this time towards the South-east Asian countries such as Vietnam and Myanmar, etc. There are many factors that influence the rise and fall of the bamboo industry in a certain country or region, and if there are no new measures to inject fresh impetus, the industry can very well go into decay. Internationally, Japan and Taiwan were the earliest to industrialize, and they

Fig. 3.114. A street view of Qingchen, Sichuan Province, China, in the 1990s (Zhu Zhaohua).

Figs. 3.115–3.117. Street views of Qingchen, Sichuan Province, China, in the 1990s (Zhu Zhaohua).

Figs. 3.115–3.117. Continued.

experienced a rise to prosperity and later gradually declined. At present, they have come to their lowest ebb. We will discuss this further in detail later in this section. So we must keep track of product and industry development trends – this way, when faced with hardship at some point, the challenges of changes in objective conditions can be met with robustness and vitality.

The inevitability and occurrence of life cycle phenomena

There are many factors that influence the path of the life cycle of bamboo products. Some of the most common elements are outlined below.

THE SUPPLY OF RAW MATERIALS. Undersupply or instability will negatively affect the development of the bamboo industry. If the price of raw materials is excessive, in most cases the removal of the products from the market or a change in production area will decrease the additional value. The example of Ecuador's bamboo industry described in Section 3.1.3 also indicates that if the industry fails to provide a stable source of raw materials, along with low prices, this will strongly

affect the bamboo farmers' enthusiasm and ultimately affect the stability of raw materials.

CHANGE IN THE LABOUR MARKET. Having the manpower with the appropriate skills and high efficiency is one of the key factors in maintaining an industry's prosperity. A few years back, I received a delegation from South-east Asia. They were stunned on witnessing the high efficiency of the workers in Anji's bamboo processing industries. They said that a single Chinese worker's efficiency is equivalent to two to three times more than one from their own county. On the one hand, this is because the level of skills mastered by them is different; on the other, it illustrates the difference in management styles. Labour wages also play an important role in deciding whether a particular product or industry can or cannot continue to operate.

CHANGE IN GOVERNMENT POLICIES. We have already discussed the local governments of Anji and Lin'an, where when a conflict in the bamboo or bamboo shoots industries escalates to a certain degree, the relevant policies will be introduced without delay to support farmers and industries to continue to sustainably develop while overcoming new challenges. However, there are also times where policies have had negative effects. For example, if the government decides on a strict environmental protection policy to control water use and air and noise pollution, small-sized pulp mills and bamboo charcoal factories will be severely affected. Nevertheless, strict policies on environmental protection are beneficial for promoting the energy-saving and conservation aspects of bamboo industries. Such policies can also facilitate the development of 'green and organic' products and the overall sustainable development of the industry.

CHANGES IN DOMESTIC AND OVERSEAS MARKETS. There are many reasons that changes arise to a product's market, including innovations in more practical, cheaper and higher quality goods. Due to intense competition, there is a demand of low prices, and the market competitiveness of products declines. Other reasons for market change are overproduction and oversupply, changes in the needs of consumer groups and changes in consumption concepts, etc.

NEW TECHNOLOGY AND PRODUCT RESEARCH. This phenomenon occurs frequently – old products become obsolete in comparison with new products. If the industry that produces the old products does not research and develop innovative products according to market needs in time, the competitive new products will replace the older ones.

OTHER FACTORS. There are many other factors that affect the survival of products or industries, including the conditions of the soil and access to electricity in some Asian and African countries.

3.8.2 Innovation is the driver for maintaining sustainable development in the industry

We must first of all recognize that the life cycle phenomenon of an industry is an objective reality that cannot be avoided, but if we always have a sense of urgency to keep innovating and finding new ways, new models and a new path forward in the face of emerging challenges, we can

resolve the difficulties. The industry, in return. is renewed with vitality.

To innovate, the first and foremost efforts should be to focus on business innovation, which includes various examples of researching and developing new technologies, products and equipment. Also, we need pioneering and innovative sales models in the new markets to strengthen and reform the management system through mobilizing the enthusiasm of employees and constantly improving staff quality and business efficiency. It is also suggested that the industrial supply chain and value chain are continuously extended, and essential to keep improving the utilization of bamboo and increasing the added value (Chen Jianyin and Zhu Zhaohua, 2005).

Local governments and communities should develop new measures and policies based on the needs and challenges for sound development of the industry and the creation of a good environment. We have already discussed in detail the bamboo development process in Anji and Lin'an. The local government assisted forest reforms, financing, taxation, demonstration bases, the training of personnel, forest road construction and the promotion of bamboo ecological and cultural industries and enterprises. A win–win partnership should be determined between communities and the masses. By taking a series of reforms and innovations, we continue to inject new impetus into industrial development. This way, the bamboo industry can sustainably develop in the long term.

3.8.3 History and development of the bamboo industry in Taiwan and Japan

Japan and China's Taiwan were among the first regions to achieve modern industrial processing of bamboo. The 1960s and 1970s were the climax of Japan's bamboo industry development. Taiwan was at its peak from the 1970s through to the 1980s, but later on, for various reasons, it started to decline in scale. At present, in both locations there are personnel who still insist on explorations of new paths of development in their bamboo industries, so in some ways there is already a ray of hope, though both are not completely out of the doldrums. Learning lessons from their development models is useful for us.

Case study 1: Taiwan's development of its bamboo industries

From June to November 2012, Taiwan's Cultural Department hosted an investigation for which CHIC BUSINESS & BRAND INC. (Cbbic) was responsible. This investigation of the status of Taiwan's bamboo industries was very comprehensive and thorough, the purpose being to understand all of the problems of the bamboo industry and the difficulties encountered with the decline of the industries. Cbbic provided bamboo farmers, businesses and relevant governmental bodies with information and advice for future development. Additionally, it also gave guidance to various levels of government on how they should strengthen support for microenterprise entrepreneurship. The 90-page survey report produced by Cbbic provided detailed information for us to understand the state of Taiwan's bamboo industry (CHIC BUSINESS & BRAND INC., 2012). Here is a very brief summary of the report.

INDUSTRY DEVELOPMENT HISTORY. The development of the bamboo industry in Taiwan can be divided into six different stages (see CHIC BUSINESS & BRAND INC., 2012; Cultural Affairs Bureau of Nantou County, 2014).

The Japanese occupation period (before 1945). During this period, Taiwan exported some traditional products to the USA and bamboo fan frames and trays to Japan. Zhang Heshun, a bamboo industrialist in Taiwan, invented a machine that manufactured chopsticks in the year 1931. This improved the processing efficiency, and therefore Taiwan exported over 7500 pairs of chopsticks a year from then. In order to meet the needs of the Japanese for bamboo products, Japan invited Taiwanese companies to teach them skills and train a group of technicians. The Japanese also established a research institute in Taiwan, which lay a good foundation for the development of bamboo industries in Taiwan.

Agricultural period (1945 to 1960). During World War II, the demand for bamboo products increased in the international market. More than 20,000 people became employed in the process of exporting bamboo products and crafts based on bamboo. In 1950, a town called Zhushan established two production cooperatives, which led to bamboo farmers increasing their overall sales and bamboo being comprehensively managed. Under the guidance of the Council of Agriculture, the construction of factories, bamboo segmentation processing, the raw material drying process and demand for rural supplies and tools rapidly increased.

The transition period (1960 to 1970). By this period, a large number of bamboo products had been replaced by plastic products, including traditional products such as chopsticks. At that time, the government actively promoted industrial development and bamboo processing machinery imported from Japan, and replaced the manual mode of production to enhance competitiveness and achieve a successful transition. The main products were: bamboo coasters, mats, upholstery, chairs, curtains, coffee trays, fountains and daily personal hygiene facilities, etc.

The formative years (1971 to 1975). During this period, the Japanese domestic production environment faced changes and wages increased. Consequently, many Japanese bamboo manufacturers went to Taiwan to seek cooperation and investment and transfer technologies in factories, thus promoting stable growth and an increase in export performance. Governments also made efforts to assist in the development of the bamboo industry, such as research on and development of bamboo processing machines, setting up training and organizing studies to enhance the industry's technology. They set up seven research institutes in several different places. In 1973, a bamboo processing industry zone was established in Zhushan, in which numerous technical personnel were trained and the bamboo industry showed booming growth. During this period, bamboo swords, toothpicks, skewers, mats, furniture, magazine racks, chopsticks, etc. were exported.

The heyday period (1975 to 1985). The government helped to set up bamboo processing industries through loans for purchasing machinery and equipment, regulating operations, researching and cooperating to encourage experiments in forestry related to the physical processing of bamboo, and anti-corrosion and coating technology research work.

The research and development of laminated board technology was a success and the value of bamboo was greatly enhanced with a good market response. Foreign markets also showed great interest, especially in furniture. Under full funding and with technology upgrading, the bamboo processing and export values grew substantially. Bamboo chopsticks, lighting, birdcages, furniture, swords, tea trays and bamboo spoons were popular.

The period after 1986. With economic development and changing lifestyles, bamboo products have been gradually replaced by other products. On top of that, rising land and labour wages caused labour-intensive bamboo processing industries to lose their competitive edge. Many industrialists have moved to Mainland China and South-east Asian countries because of their low wages and the ability to use factories producing bamboo products to sell back to Taiwan. This led to a loss of manufacturers and employees, and a decrease in the gross value of production. The number of bamboo processing industries went from 1500 to less than 500 in 1993 and still lower to 100 in 2004.

TAIWAN'S CURRENT SITUATION.   After 1986, due to the reform in Mainland China, China became rich in resources and had very cheap labour costs. Bamboo entrepreneurs from Taiwan sold large amounts of processing equipment to Mainland China. From 1986 to early 1990s, the bamboo industry in Taiwan played a great role in promoting Mainland China's bamboo sector, industrial processing, product development, access to international markets and other aspects. Taiwanese entrepreneurs derived a lot of benefits, but Taiwan itself suffered a major blow. Although there are still some small businesses on the island, because local wages are too high, the brain drain phenomenon, the local market being limited, and several other reasons, the bamboo industrialists in Taiwan began to recover after nearly 30 years of struggling. In 2013, there were 293 bamboo industries in Taiwan, which included 66 raw material industries, 152 processing industries and 75 marketing industries. Overall though, the bamboo industry in Taiwan has yet to shake off the doldrums. The main changes that have occurred in the recent period are described below.

Bamboo supply instability and farmers' frustrations. According to 2012 statistics from Taiwan's Council of Agriculture, the country's bamboo forest area was 149,516 ha, accounting for 11% of Taiwan's forest area. Due to the shrinkage in scale of Taiwan's bamboo processing industry, the reduction in demand for bamboo and decrease in prices coupled with the impact of imports, only 55% of industries used local bamboo as raw material. The rest used imports from Mainland China and Vietnam. Taiwanese bamboo production greatly shrunk from 17,870,000 culms in 1976 to 574,032 in 2004, which is not even 5%; in 2012, the number improved to 1,741,762. Taiwan's bamboo farmers are not as interested in operating bamboo industries, and have changed their focus to tea, where the profits are far higher than for bamboo. Another reason for the decline of Taiwan's bamboo industry is the ageing of the bamboo farmers, who are no longer able to carry out intensive management of their bamboo stands. Also, the time consumed in harvesting and transporting is great, and the equipment for cutting bamboo is costly. Further, Taiwan's *B. brumeana* sold for only NT$5000/ha, and the cost of workers was about NT$7000–8000/ha, so the process is not economic. Taiwan's bamboo farmers are more interested in harvesting bamboo shoots, as in Ryugasaki in Japan, the average price of bamboo shoots is NT$80–85/kg, and there is an annual output of 500 tons. So farmers are interested in harvesting bamboo shoots and tea leaves.

Downsizing of the bamboo processing industry. Taiwan's bamboo processing enterprises are mostly small family enterprises, the average annual turnover for each family being around NT$9.4 million, with a sales income of NT$5.01–8.0 million. Approximately 50% of processing industries are in the growth stage, with products such as traditional bamboo art, charcoal, fabric, flooring and other building materials; 20% of the industries remain unchanged; and 25% have negative growth with products such as gardening supplies, household items, raw materials and bedding supplies. Better market prospects can be seen from the current demand for bamboo products and art. The reason that Taiwan's midstream processing industry is experiencing difficulties is: production costs, narrow sales channels, insufficient competitiveness of products (prices and R&D), and an incomplete supply chain and value structure, although most hold a positive attitude. Some 55% of industries say they want 'to accelerate industrial upgrading and transformation and innovation policies' that advocate and strive to develop new products and business alliances, hire foreign workers and use other measures to achieve industrial upgrading. This shows that even though Taiwan's bamboo industry faces difficulties, many are still confident in the future, which is a valuable driving force in the future of Taiwan's sustainable growth.

Domestic and overseas sales market expansion. From the sales point of view, approximately 65% of Taiwanese industries have both domestic and overseas markets (20% operate only in the domestic market and 10% are overseas oriented industries). Marketing methods have a great impact on the business situation of enterprises. All of the companies who have both domestic and overseas markets are generally in a state of gradual increase and expansion, but enterprises that have only domestic markets usually show negative growth, which indicates that the size of Taiwan Island makes the markets limited. According to the survey by Cbbic in 2012 (CHIC BUSINESS & BRAND INC., 2012), the size of the Taiwanese bamboo products market is about NT$4.705 billion, or about US$152 million, so its international market must expand. In 2012, Taiwan's bamboo products export value was more than US$37 million, so there is a gradual recovery of Taiwan's bamboo industry.

TAIWAN'S BAMBOO INDUSTRY IS PREPARING TO ACCUMULATE ENERGY FOR FUTURE DEVELOPMENT. Taiwanese industrialists think that although the market is small, because bamboo products are durable, nice to touch and have natural environmental features, many consumers will still continue to demand bamboo products such as household items and furniture. The industry can dictate its future direction, as some areas of products have prospects, such as the use of bamboo for garden landscaping, in addition to cultural and creative daily necessities, bamboo materials and furniture. Bamboo is known as 'new green energy'. The Taiwan Institute of Technology proposed the idea of 'green earth bamboo craft' as a green

economy concept. In recent years, Taiwan's research institutes have closely cooperated with the bamboo industries to develop a series of new competitive products. These products include high quality bamboo charcoal, the use of highly elastic non-woven bamboo to make hats, scarves, gloves, socks, laundry shampoo, insect repellents and roasted peanuts (bamboo charcoal peanuts), as well as devices used to purify water and air. In 2008, the Taiwan Institute of Technology cooperated with the Creative Design Center to exhibit a series of works called 'yii', in Paris and other parts of Europe. A bamboo stool won an award for people's choice of favourite bamboo product (Lin Yinggui and Zhu Gezheng, 2010; Cultural Affairs Bureau of Nantou County, 2014; Xiao Aijun, 2014).

Changes in Taiwan's bamboo industry are embodied by Zhushan Township in Nantou County (central Taiwan) (Xiao Aijun, 2014). Here was the centre of Taiwan's bamboo industry in the 1970s, when it had hundreds of bamboo factories, though now there are only 50 such businesses. On 21 September 1999, Zhushan experienced a strong earthquake, and after that, the government and business communities decided to revitalize the old industries and the town began to recover. The Taiwanese government set up the Taiwan Industrial Technology Research Institute in Zhushan to provide assistance for local economic development by finding a way to revitalize the bamboo industry. The Institute has developed a series of new products, such as bamboo charcoal products, flooring, crafts pens, laptop jackets and so on. The diversified development of the bamboo industries brought new hope for the town. Zhushan town has also built a 'Bamboo Culture Park', a combination of tourism, leisure, nature conservation, local industries and an education base for understanding bamboo culture. It can be seen that Zhushan's bamboo industry has not repeated its old business models, but in the face of new challenges, it has not only developed products that are high tech, high added value and more sophisticated, but also combined ecology, tourism, culture and education all together in a higher level of industrial development for the future.

What is gratifying is on 18 August 2013, an NGO related to bamboo was established – the Taiwan Bamboo Society (中文是台湾竹会), whose members include researchers in the field of bamboo, architects and engineers, craft artists and bamboo cultivation and nursery experts (Tang Lifang, 2010). The association is a platform for promoting information, technology, culture and education about bamboo and it strives to build a link between primary, secondary and tertiary industries. Those who are passionate about bamboo and nature have jointly promoted Taiwan's bamboo culture and innovations. Here we would like to offer hearty congratulations to the Taiwan Bamboo Association for its efforts.

Case study 2: the evolution of Japan's bamboo industry

(The contents of this section are mainly extracted from the presentations of Prof. Shozo Shibata to the 10th World Bamboo Congress in Damyang, South Korea, see Shibata, 2015a,b and the thoughts of the two authors).

Japan has a long history in the development of bamboo and bamboo culture has a strong presence. The country is a pioneer in bamboo processing within the industrialized countries. Japan also had a major impact on the early development of Taiwan's bamboo industry. It has a bamboo forest area of 160,000 ha (60% is concentrated in Kyushu and 97% of the bamboo forests are privately owned), with an annual output of 20–30 million tons of bamboo. Popular products include daily necessities, clothes hangers, decorations and fences. Currently, most of the bamboo enterprises are small factories owned by families, with about 100,000 workers employed in this way.

Japan's bamboo industry reached its peak from 1960 to 1970, but due to production costs, it then began to decline. After the mid-1980s, the country began to import large quantities of bamboo products, especially from China, and a large number of processing technologies began to shift to China (Cultural Affairs Bureau of Nantou County, 2014).

This period resulted in the weakening of Japanese bamboo industries. As seen from Fig. 3.118, the 128,000 ha of managed bamboo forests in 1980 decreased to 38,000 ha in 2009. In the absence of management, bamboo scattered through free expansion and so the actual total area of bamboo forests increased. The area in 1980 was 144,000 ha and it increased to 160,000 ha later. This phenomenon

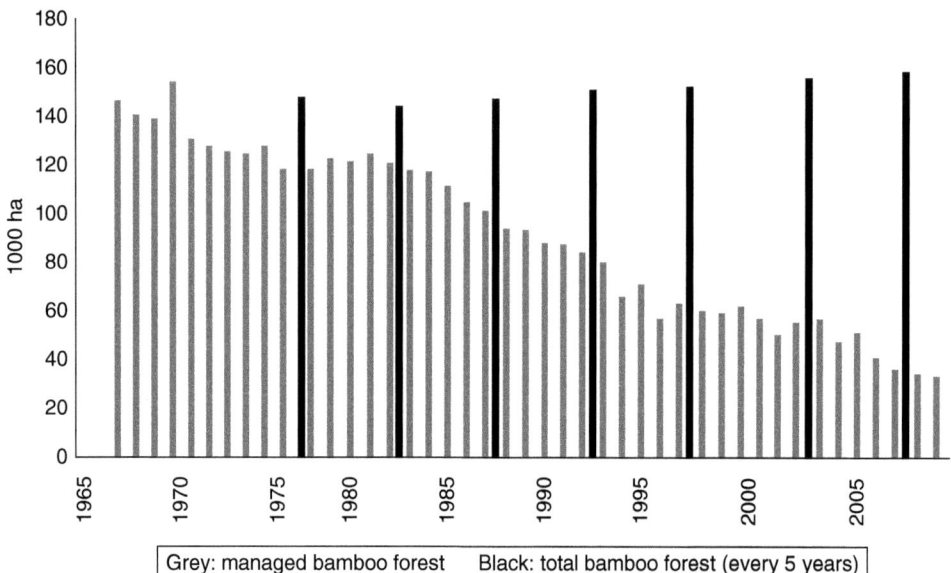

Fig. 3.118. During the period from 1960 to 2010, the area of managed forest (grey bars) in Japan decreased, and bamboo production varied greatly. The output from *Phyllostachys pubescens*, for example, fell from an annual amount of 180,000 culms in 1980 to 25,000 culms in 2010 (data from the Japan Forestry Agency, 2015; modified from Shibata, 2015a,b).

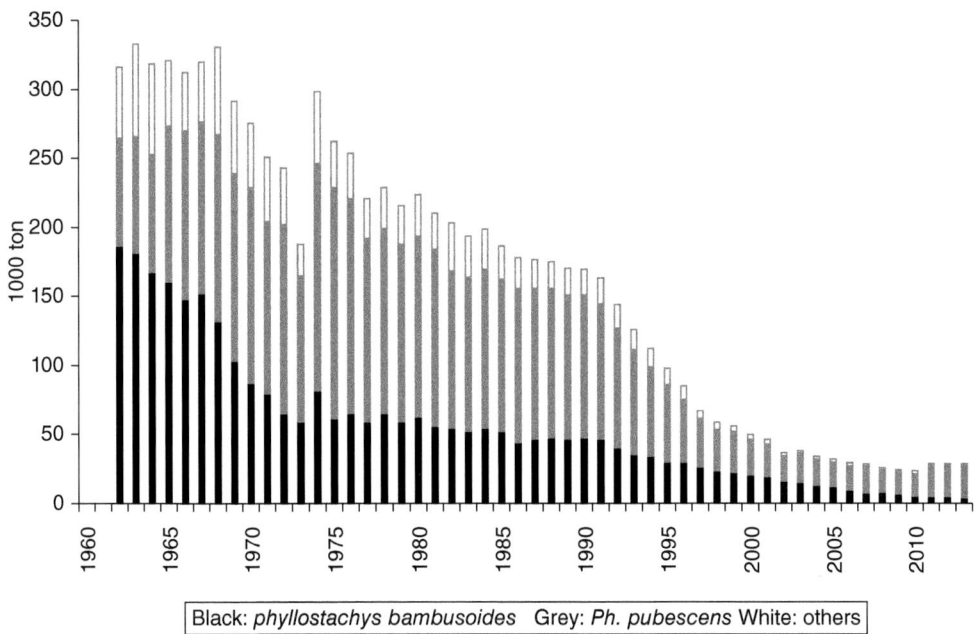

Black: *phyllostachys bambusoides*   Grey: *Ph. pubescens* White: others

Fig. 3.119. The output of 'timber' from *Ph. bambusoides*, *Ph. pubescens* and other bamboo species in Japan from 1960 to 2010 (data from the Japan Forestry Agency, 2015; modified from Shibata, 2015a,b).

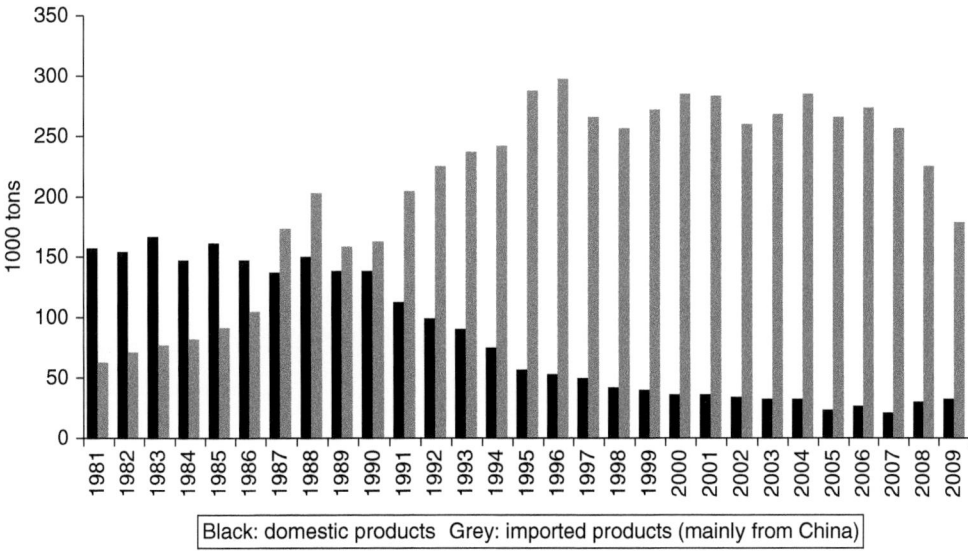

Black: domestic products   Grey: imported products (mainly from China)

Fig. 3.120. Comparison of the amounts of bamboo shoot products exported out of and imported into Japan (mainly from China) from 1981 to 2009 (data from the Japan Forestry Agency, 2015; modified from Shibata, 2015a).

of free expansion of bamboo had once aroused discontent from some local Japanese residents, because the forests 'invaded' their crop farms and ranches.

Shoot yield also declined significantly, with the amount of boiled bamboo falling from 150,000 tons in 1983 to 26,000 tons in 2009. At the same time, imports from abroad went up from 70,000 tons in 1983 to 180,000 tons in 2009; of this, most of the imports were in the time period of 1995 to 2004, with an average of 26 million tons. In 1996, there were up to 30 million tons imported.

Because of Japan's rapid industrial development from 1960 to 1980, the country emerged as the world's largest economy, so the cost of labour increased greatly, and hence labour-intensive industries such as the bamboo industry began to decline (see Fig. 3.120). During this time, Taiwan's bamboo industry did well in the market and achieved its peak performance. Starting from the 1980s, due to the rapid development of bamboo industries in Mainland China, a large number of low-cost and high-quality bamboo products entered the Japanese market from China, resulting in a further decline of Japan's bamboo industry.

Although Japan's bamboo industry has suffered such a blow, because the country has a long bamboo cultural tradition, the people still like bamboo and its products. They also like to eat bamboo shoots, and Japan has the world's highest per capita sales of bamboo shoots. Japan also has a strong scientific and technological power, so has tirelessly done research on bamboo and developed new technologies and high value-added products. For example:

1. Japan was the first country to develop bamboo charcoal, bamboo vinegar, bamboo deodorants and health-care products, and bamboo cosmetics, soil conditioners and the like. Toshiba also uses bamboo fibre, charcoal and the opal inside bamboo leaves to make microphones.
2. The outer layer of laminated bamboo material is used as an antibacterial.
3. Japan leads the development of bamboo fibre, which is used for paper, clothing, food and now for growing nanoparticles such as bioplastics and biogas. Shoes are also made with bamboo and fibreglass (see Fig. 3.122).
4. Bamboo leaf extracts are used for medicine, fungicides and cosmetics.
5. Bamboo is used for making biomass particles and biogas.
6. Japan has also made new developments in the machinery and equipment for bamboo processing, production and harvesting, including the making of bamboo fibre biogas and powder particles.

At the same time, as new technologies and products are developed, Japanese academics, its

Fig. 3.121. A bamboo fibre factory in Japan (from Shibata, 2015b).

Fig. 3.122. Shoes made from bamboo and glass fibres in Japan (from Shibata, 2015b).

business community and the public are now reflecting upon the process of development of the bamboo industry. It is essential to see the value of bamboo for the environment and this should be subject to detailed assessment, restoration, revegetation, biodiversity conservation and landscaping.

CURRENT SITUATION. Japan has also continued efforts to produce traditional handicraft products under the new conditions, but in recent years, the country has been trying to develop new bamboo products. Japanese artists are paying more attention to art made from bamboo, and some works have been exhibited in art galleries and museum collections. The Japanese seek refinement, so they have a strict selection process for bamboo. They are aiming for subtlety and quality over quantity (CHIC BUSINESS & BRAND INC., 2012). The Ministry of Agriculture and Forestry is responsible for the bamboo industry in Japan, and the country has many societies for promoting bamboo, bamboo culture, the bamboo industry and bamboo charcoal. Japan hosts a Bamboo Culture Festival each year. We look forward to Japan's bamboo industry continuing to grow from strength to strength.

Case study 3: the way to develop – continual innovation – the case of Hangzhou Dasso Co., Ltd.

Every transformation of an enterprise is a painful process, but it is necessary to make transformations in a timely manner in accordance with market demand and changes, so that enterprises can survive and develop. Each transition will mean new capital investment and also the creation of new technology. New market development and the emergence of high value-added products give space for progress and development.

Here, China's largest bamboo panel enterprise, Dasso, has provided us with a success story (Dasso, 2014). Since 1993, the company has been processing bamboo by closely following the development trends of the international market (strengthening the development of new products and technologies, investment, close cooperation with domestic and foreign research institutes and experts, etc.). The company continues to amaze people by its new uses for bamboo. In 2003, it first introduced a slicing technology for veneer that sped up large-scale production of bamboo veneer. In 2005, Dasso successfully developed fireproof panels made of bamboo timber materials. In 2006, Cambridge University and Dasso jointly developed a wind turbine project,

Fig. 3.123. Madrid Barajas International Airport in Spain showing the bamboo ceiling made of planks of multi-layered Moso bamboo veneer (photo by Dasso).

Fig. 3.124. Bamboo-based wind turbine blades (photo by Dasso).

Fig. 3.125. Shandong Grand Theatre in Jinan, Shandong Province, China (photo by Dasso).

Fig. 3.126. Wuxi Grand Theatre in Jiangsu Province, China (photo by Dasso).

followed by offering BMW green materials for the interior decoration of its X5 and X6 sports utility vehicles (SUVs) in 2007. Dasso also completed the decoration of Madrid Airport in 2007. In 2008, while collaborating with Nanjing Forestry University, the company researched and developed outdoor bamboo flooring that set the standards for the country. In 2009, Dasso completed 100 sets of wind turbine blades (800 kW) and these operated successfully; the company also launched the industrialization of new material technology.

Dasso has also contracted a serious of major construction projects. In 2009, it completed a large-scale indoor and outdoor construction project called Vanke Metropolis in Beijing. The company also completed the renovation of the Wuxi Grand Theatre in Jiangsu Province 2012 and satisfied the acoustic requirements for a large concert hall. Shuijingfang Museum in Chengdu, Sichuan Province, Shandong Grand Theatre in Jinan, Shandong Province, and the deck in Qiandao Lake in Hangzhou, Zhejiang Province, were also some of Dasso's accomplishments.

At the same time, the company has developed a series of home decoration materials, structural materials made of bamboo, firewood, outdoor furniture, wood preservatives, high-strength engineering materials, interior decoration materials and so on. It is not difficult to comprehend why Dasso has become China's largest enterprise and has such a great influence in the international bamboo industry. Because of its persistence in the exploration of high-tech bamboo products, and ambition in undertaking projects with large impacts, the enterprise has achieved great growth in strength and has entered a sustainable development period.

Market competition is brutal. We have described some examples of successful transitions. However, due to inadequate market research, inadequate investments in new product development, chaotic competition and the impacts of the international financial crisis, some bamboo companies have failed. Since 2012 especially, in addition to the above factors, the rapid increase in labour wages and strict factory governmental requirements have caused yet more businesses to fail.

Note

[1] The phrase 'well-off society', designated 'moderately prosperous society' (Xiao Kang She Hui) by Wikipedia, was first used in the *Classic of Poetry* (*Shijing*, 《詩 經》) written as early as 3000 years ago. 'Xiaokang' may be associated with an Engel's coefficient of 40–50%. The Chinese leader Deng Xiaoping used the term 'Xiaokang society' in 1979 as the eventual goal of Chinese modernization. In the usage (tifa) of current General Secretary Xi Jinping, the term 'China Dream' or 'Chinese Dream' has gained somewhat greater prominence. During the annual National Party Congress meeting of 2015, Xi Jinping also unveiled a set of political slogans called the 'Four Comprehensives', which include 'Comprehensively build a moderately prosperous society' (sourced from Wikipedia entries).

References

Many of the references included here have only minimum details and some are informal papers that have been produced for training courses, assessment visits to other countries, etc., but they are included in order to give some indication of the sources used for this chapter.

Anji Forestry Bureau (2002) *The In-Depth Development of the Bamboo Sector – the Three Leaps Achieved by the Anji Bamboo Sector* (发展竹产业 做深竹文章——安吉竹产业发展实现三个跨越). Anji, China.
Anji Forestry Bureau (2013a) *Basic Facts of Anji Bamboo Sector* (安吉县竹业概况). Anji, China.
Anji Forestry Bureau (2013b) *Brief Introduction to Anji County 2012* (2012年安吉县情况简介). Anji, China.
Anji Forestry Bureau (2015) *2014 Year Book of Anji Statistics* (安吉统计年鉴). Anji, China.
Chen Jianyin and Xuan Taotao (2007) Bamboo Farm Stay in Anji. Anji, China.
Chen Jianyin and Yan Guoqin (2002) Anji – the Drivers of Bamboo Industry in Other Bamboo Producing Areas of China. Anji, China.

Chen Jianyin and Zhu Zhaohua (2005) Increasing Bamboo Utilization Rate and Added Values. Anji, China.

CHIC BUSINESS & BRAND INC. (2012) *Basic Investigation on Taiwan Bamboo Sector and the Researches on Bamboo Materials* (台湾竹产业基础调查与竹材研发调查). CHIC BUSINESS & BRAND INC., Taipei.

Cultural Affairs Bureau of Nantou County (2014) Evolution of Taiwan Bamboo Sector (台湾竹产业的演变). Available at: http://www.nthcc.gov.tw/bamboo/ (accessed 4 April 2017).

Dasso (2014) World's Trend of Bamboo Utilization. Hangzhou Dasso Co., Ltd., Hangzhou, China.

Judziewicz, E.J., Clark, L.G., Londoño, X. and Stern, M.J. (1999) *American Bamboos*. Smithsonian Institution Press, Washington, DC.

Lin'an Forestry Bureau (2003) Master Plan of Lin'an Forestry in the Thirteenth Five Year Plan (临安市十三五林业总体规划). Lin'an, China.

Lin'an Forestry Bureau (2014) 2014 Yearbook of Lin'an Statistics (临安统计年鉴). Lin'an, China.

Lin Yinggui and Zhu Gezheng (2010) Investigation on Taiwan Bamboo Art Culture and Bamboo Material Researches (台湾竹艺文化传承与竹材研发调查). Taipei, Republic of China. Available at: 163.17.30.196/blog/lib/read_attach.php?id=30464 (accessed 19 July 2017).

PI (2010) Investment Opportunity – Bamboo Mat Board Prospectus for Hanoi, Vietnam. Prosperity Initiative, Rapid City, South Dakota.

Shibata, S. (2015a) Bamboo resources for new usage in Japan. Paper presented under the Theme: Ecology and Environmental Concerns. In: *10th World Bamboo Congress Proceedings, 17–22 September, 2015, Damyang,* Korea. World Bamboo, Plymouth, Massachusetts. Available at: http://www.worldbamboo.net/wbcx/Sessions/Theme%20Ecology%20Environmental%20Concerns/Shibata,%20Shozo.pdf (accessed 26 June 2017).

Shibata, S. (2015b) Bamboo resources for new usage in Japan. PowerPoint presentation under the Theme: Ecology and Environmental Concerns. In: *10th World Bamboo Congress Proceedings, 17–22 September, 2015, Damyang, Korea.* Available at: http://www.worldbamboo.net/wbcx/Powerpoints/Ecology%20and%20Environment%20Concerns/Bamboo%20Resources%20for%20new%20usage%20in%20Japan_ShozoShibata.pdf (accessed 27 June 2017).

Sta. Ana, R.L. (2015) Impact of MOST/ INBAR Workshop. Manila, the Philippines.

SFA (2013) *Development Plan for China's Bamboo Industries (2013–2020)* (全国竹产业发展规划). State Forestry Administration, Beijing, China.

Tang Lifang (2010) Introduction to Taiwan Bamboo Society (台湾竹会介绍). Taiwan Bamboo Society, Taipei.

Wang Anguo (2005) Linking the Government, Enterprise and Farmer: Bamboo Industry Society of Lin'an, Lin'an, China.

Wang Anguo (2009) Multi-Participation in the Bamboo Shoot Industry (临安竹笋产业发展的多方参与). Lin'an, China.

Wang Anguo (2014) Review of Lin' an Bamboo Shoot Industry Development. Hangzhou, China.

Xiao Aijun (2014) How to Revive the Bamboo Sector in Taiwan (台湾竹业如何能浴火重生). Available at: http://www.bbc.com/zhongwen/trad/fooc/2014/04/140414_fooc_tw_bamboo_booming (accessed 14 April 2014).

Yang Yuming (2000) *A Survey Report on the Bamboo Species in Ecuador*. Guayaquil, Ecuador.

Zhu Zhaohua (2000) A Report on Study Tour to South America. Beijing, China.

Zhu Zhaohua (2001) A Report on My Visit to Three Countries in Africa. Beijing, China.

Zhu Zhaohua (2002) Further Suggestion to Ecuador's Bamboo Industry Development. Quito, Ecuador.

Zhu Zhaohua (2005a) A Report on My Visit to Three Countries in Latin America. Beijing, China.

Zhu Zhaohua (2005b) A Report on the Machinery Expert- Mr. Wu Qing's Consultation in Ecuador. Beijing, China.

Zhu Zhaohua (2005c) *Bamboo Garden of Grupo Wong in Santo Domingo*. Guayaquil, Ecuador.

Zhu Zhaohua (2005d) Report on the Visit to Colombia.

Zhu Zhaohua (2006) Report on the Consultancy for the Bamboo Factory, Vietnam. Hanoi, Vietnam.

Zhu Zhaohua (2007) Evaluation of the Bamboo Industry's Impact on Rural Sustainable Development in Anji, China. Presentation by INBAR (International Network for Bamboo and Rattan, now the International Bamboo and Rattan Organization, Beijing) to an International Training Workshop on Non-Timber-Forest-Product (NTFPs) Development and Utilization in Zhejiang, China.

Zhu Zhaohua (2007a) A Case Study on Successful Sustainable Management of Natural Forest Resource in Baisha Village. Lin'an, China.

Zhu Zhaohua (2007b) Impact Assessment of the Bamboo Shoot Sector on Poverty Reduction in Lin'an. Lin'an, China.

Zhu Zhaohua (2010a) A Report on the Study Tour of the Chinese Experts for MB Projects. Hanoi, Vietnam.

Zhu Zhaohua (2010b) Decision Encouragement Policies for the Bamboo and Rattan Sector. Hanoi, Vietnam.

Zhu Zhaohua (2011a) Analysis of the integrated sustainable development of Lin'an – economy, ecology and livelihood. [PowerPoint version of] Plenary Session Keynote Speech to: *International Conference, Multipurpose Forest Ecosystem Management in a Changing Environment, 23–25 November, 2011, Nanning, Guangxi, China*. Available at: http://www.valwood.uni-freiburg.de/Downloads/Nanning/KEYNOTES/Zhuzhaohua-Analysis%20of%20the%20Integrated%20Sustainable%20Development%20of%20Lin%E2%80%99an.pdf (accessed 30 June 2017).

Zhu Zhaohua (2011b) [Plenary Session Keynote Speech:] A successful case of integrated sustainable development of [a] mountainous area – analysis of the economy, ecology and livelihoods development in Lin'an. In: Book of Abstracts, [*International Conference,*] *Multipurpose Forest Ecosystem Management in a Changing Environment, 23–25 November, 2011, Nanning, Guangxi, China*, pp. 42–58. Available at: http://www.valwood.uni-freiburg.de/Downloads/Nanning/Book%20of%20Abstracts.pdf (accessed 30 June 2017).

Zhu Zhaohua (2012a) Analysis of the integrated sustainable development of Lin'an – economy, ecology and livelihood. In: Tewari, D.N. (ed.) *Forestry for Sustainability*. Ocean Books, New Delhi.

Zhu Zhaohua (2012b) Full Utilization of Bamboo in China. Anji, China.

Zhu Zhaohua (2012c) *Visit to the Philippines*. Baguio, Philippines.

Zhu Zhaohua and Chen Jianyin (2013) The Sustainable Development of Anji's Bamboo Industry. Anji, China.

Zhu Zhaohua and Jin Wei (2006) The Supply Chain of Bamboo Industry and Full Utilization of Bamboo Materials. Armenia, Colombia.

Zhu Zhaohua and Wang Anguo (2014) Basic Information on Mountain Areas Sustainable Development. Hangzhou, China.

4

Develop or Create a Featured Bamboo Industry According to Local Conditions

———————————

By reviewing the various factors affecting the sustainability of the bamboo sector, it can be concluded that the sector is complicated and that its sustainability is subject to various requirements/conditions:

1. The first is the creation of a comprehensive strategic plan with many aspects of the local ecology, economy and society considered.
2. Another indispensable condition is a close cooperation among the stakeholders – the farmers, enterprises, government and scientists.
3. The third condition is to keep tight connections between the primary, secondary and tertiary industries of the sector, so that the cultivation of the bamboo resource, its processing and its marketing form a complete supply chain.
4. The fourth is to be creative in the face of various difficulties in the course of development, the changes and challenges to be met, and the new methods that need to be innovated to take the sector into a higher level of development.
5. The fifth is to create a multi-participatory cooperation mechanism that will realize a common share of the benefits and provide a multi-win situation.

4.1 Identify a Strategic Plan Suitable for Local Conditions and Characteristics

4.1.1 What to do with the successful experiences from other countries?

In the earlier parts of the book, we discussed the many successful and failed cases in the bamboo sector. China's successful experiences in bamboo sector development in particular were introduced in detail.

I (herein after refer to the first Author – Prof. Zhu Zhaohua) believe the first thing to do is to understand the universal guiding significance of these experiences: that when conducting bamboo development, one should give special attention to the key factors that affect the sustainability of the bamboo sector. Although the priorities in different development stages can be different, these factors need attention all the time. The experiences of success and failure, and the models accumulated by China and other countries all provide valuable references for a bamboo developer.

However, when recognizing the universal guiding significance of these experiences, we should also pay attention to their particularity.

This is because the experiences came from a specific country or region and had specific conditions. For example, the product structure would have been designed according to the local resources, personnel and technology that had been accumulated, and according to domestic and international market trends; in other words, the policies identified were based on the problems that emerged under certain local conditions. When we are talking about land policies, China's land user rights policy may not apply to other specific regions. This is because, in China, all of the land and the resources were state owned, not privately owned, although the management rights could be state owned, collectively owned or privately owned. In many countries of the world, most lands are privately owned, and ownership of the bamboo forests should be resolved as a first and necessary step in development of the bamboo sector. At present, in many countries in Africa, Asia and Latin America, the ownership of the natural bamboo forests is not clear, and this may hinder the rehabilitation and intensified management of the forests, and discourage the local communities from participating. Hence, the identification of bamboo forest management rights is a must for developing any bamboo sector, and while local situations may be different from those in China, management changes can still be made by applying different models from those used in China.

To sum up, when studying the experience of bamboo sector development from other countries, one should neither refuse to learn just because the conditions are different – because those experiences may have some universal significance, nor copy everything exactly without due analysis and modification.

4.1.2 Identify/innovate development strategies suitable for local conditions and features

Like other sectors, the bamboo sector of a certain area, country or region will have its own features and conditions. Blind investment without studying these conditions and features will result in detours or failures. Besides local conditions and features, understanding the status and trends of bamboo development globally is also necessary. When identifying a strategic plan for a bamboo sector, one should combine the most recent results of the world bamboo sector with the local conditions, and explore the best suitable strategy.

Here let us start by reviewing again the long history of bamboo development in Anji, Zhejiang Province, as already discussed in Chapter 3 (Sections 3.3, 3.5.2 and 3.7). Before the industrial processing of bamboo was developed, Anji already had a strong traditional sector, and there were already several series of products, such as bamboo woven products, carved products, furniture, pulp and construction materials. There was a long period of manual processing of bamboo in local areas before machines were introduced from Taiwan in 1986. The mechanical processing of bamboo, started in a specific period of China's social, economic and political development, has developed fast from a micro or small scale to large scale in the period 1985–1995. The number of processing factories in Anji increased from 30 in 1985 to 300 in 1995. Since 1995, some state-owned factories have started to transfer private ownership. Up to 1998, the total number of factories in Anji reached 1500. At the same time, because of capital accumulation, the scales of the factories were gradually enlarging, and some of the homestead factories had been developing into medium- or large-scale modern enterprises.

Up to 2000, a number of medium- and large-scale enterprises became the backbone of Anji's bamboo sector, and led a large number of small enterprises or microenterprises processing semi-products. A supply chain from raw material supply to primary processing, to semi-product processing and then finalizing, came into being. This supply chain had a comparatively finer and more professional division of work, which resulted in higher production efficiency, lower costs and better quality products. At this period, although the enterprises were independent in their management, they were also partners in one supply chain. Up to the year 2010, Anji had already formed an industrial group system, with more than 3000 types of bamboo products in nine categories, 2300 enterprises.

After 2010, in face of the challenges brought by higher labour and raw material costs, and the lack of innovation capacity and competitiveness, the local government started to plan the construction of a bamboo industry park, which would gather the most competitive enterprises in Anji, and introduce cooperation with related research institutes and universities. Later, the park developed comprehensive industries that combined real production with the ecological and cultural industries, and combined enterprises with research and innovation groups. The comprehensive development model of the bamboo park brought Anji's bamboo sector to a higher level of development, and to date, the park has just begun to take shape. It will definitely inject new power into Anji's bamboo sector.

How should Anji's development experiences be studied according to local conditions? The development course of the bamboo industry in Anji over the past 30 years (1986 to 2015) has changed from traditional to modern sectors, from micro- and small-scale to large-scale group production, from real production alone to a comprehensive sector that integrates the environmental, ecological, cultural and tourism industries. Now, Anji's sector is on a sustainable development road. The purpose of reviewing the course of development of Anji's bamboo sector is not to have other nations and regions follow in the footsteps of Anji. Rather, it is because the course of Anji's bamboo development has included various types of development models: traditional manual processing models; micro- and small-scale homestead models; group production models that aggregate large- and medium-sized backbone enterprises, and large numbers of primary processing factories and semi-product factories; as well as the (latest) comprehensive sector model that combines real production with the ecology, culture and tourism industries. Different countries and regions may select their own suitable models and cut-in points from these.

It was mentioned in Chapter 3, Section 3.8.3, that Taiwan's bamboo sector experienced its glorious period from the 1970s to 1980s, and a downturn period from 1986 to 1999. From 2000, Taiwan's bamboo industry was concentrated in Zhushan Town, Nantou County, where new bamboo enterprises emerged, the Taiwan Bamboo Industry Technology Research Institute was established, along with the 'Bamboo Cultural Park', which combined production development with bamboo sightseeing and leisure

tourism. These had pushed the development of Taiwan's bamboo sector to a higher level and enabled the recovery of the sector little by little.

The Japanese bamboo sector has experienced more dramatic changes (see Chapter 3, Section 3.8.3). From the 1960s to the 1970s, the Japanese bamboo sector reached its peak development period, but after the downturn in the 1990s, the government and the people even started to hate bamboo, and some even tried to eliminate the bamboo forests. Since then, Japan has had to import bamboo raw materials and products to replace the production of its own bamboo sector. However, under the efforts and insistence of some experts and master craftsmen, Japan has now strengthened its bamboo research, and people have rediscovered the values of bamboo – for besides its use for daily utensils and as food, bamboos can also be developed into fine and exquisite artistic products that have great added value. At the same time, Japanese researchers found the ecological and cultural values of bamboo, and started a higher level of bamboo development. The cases of Taiwan and Japan told us that although they have experienced a glorious development period, these countries and regions cannot return to their old paths, but have to explore new development paths and modes according to the ever-changing conditions and in line with new trends.

4.2 Explore the Breakthrough Point of Bamboo Sector Development

4.2.1 Factors that need to be considered while selecting the breakthrough point

To develop a bamboo sector, usually one should first answer this important question: from which product should we start? This is what we mean by the breakthrough point. Alternatively, if the development of the bamboo sector has met a bottleneck at a specific development stage, we need to find a new development mode, a new high value-adding, market-pro product. This is also what we call a breakthrough point. The correct decision on the breakthrough point may be vital for the success or failure of the whole bamboo sector. The appropriate decision making on the breakthrough point is subject to many factors covering various aspects. For example, will the locally available bamboo resources be able to meet the requirements of the new product? Is there a human resource advantage? Are the local people or relevant experts familiar with the key technologies involved? Does the new product meet the needs and demands of the local people and the social markets? Is the product permitted and encouraged by local policies? Will it bring considerable social and economic impacts to the local area, e.g. increased employment and income? What is the economic efficiency of the new product, etc.?

4.2.2 Example cases of breakthrough points

There have been various different breakthrough points in different locations. Here are three examples from China and one from Vietnam.

Anji, Zhejiang Province, China

Because of the very good technical base of manual processing over its history, once mechanical processing was introduced in Anji, a large number of microenterprises started to grow exuberantly, producing items that were mainly for daily use, such as bamboo mats, curtains, toothpicks and fans, etc.. The whole bamboo sector then gradually scaled up, and later was able to develop and produce high value-added and high-tech bamboo products that were seen as advanced in both China and the rest of the world.

Lin'an, Zhejiang Province, China

Like Anji, Lin'an had a good base of bamboo resources and traditional manual production, but chose different breakthrough points: the production of plantations for shoot purposes and shoot processing; and bamboo flooring processing as the breakthrough from culm processing. Shoot production had originally greatly increased the local rural income, but Lin'an now also became China's largest bamboo shoot production base. The highest annual fresh shoot production ever recorded was 280,000 tons. Bamboo flooring became the breakthrough point for bamboo culm processing but was very

occasional, and arose because in the 1980s, a Taiwanese machinery manufacturer wanted to invest in Lin'an in bamboo flooring production, and then established a joint venture with a local company. This first factory had led and promoted local bamboo panel production, and now Lin'an's bamboo panel processing industry has reached a considerable scale.

Xinyi, Guangdong Province, China

Xinyi County's major bamboo species is *Bambusa chungii*, a very good material for weaving. Many local farmers were skilled at bamboo weaving techniques, but most of them were weaving for household use, or for local markets. There was no scaled business nor commercialization. The breakthrough point in Xinyi was applying the model of 'company + farmer households' (see Chapter 3, Section 3.6.3), which helped in the scaling up of bamboo woven handicraft production, and reached out to the international markets. The company was responsible for providing resolutions to two problems: scaling up the production and reaching the markets. The company collected suitable products from the farmer households, then lacquered and packed them. Under this public–private cooperation, more than 2000 local households were organized for large-scale handicraft production, which in the past, had been considered hard to realize. Commercialization was also one of the advantages of the companies, as while a trade company may help to seek international customers, once the order has been placed, the production work can be divided among different households. In this way, household produced products could access international markets. Xingyi's development model made it possible for small and low-value products to develop into a large-scale industry and reach international markets. In year 2003, the production value of bamboo baskets, a major product of the county, reached US$120 million. This model realized a win–win situation for the company(ies) and the local communities.

Vietnam

Vietnam's bamboo resources are rich, and in 2006–2008, it was estimated that the total bamboo forest area could be about 1 million ha.

Fig. 4.1. Bamboo woven products being packaging by a trade company for export (Zhu Zhaohua).

In the 1980s, Vietnam attached great importance to the development of bamboo resources, and about 20,000-25,000 ha of new plantations were established, most of which could be used for industrial production. One of the major bamboo species is *Dendrocalamus barbatus,* a good material for panel processing, but there are in total more than 60 species of bamboo in Vietnam, many of which are fine for utilization; for example, fine species for bamboo woven products are *Schizostachyum funghomii* and *B. chungii.* Therefore, there was a good resource base for industrial development. The breakthrough point in Vietnam was simple industrial products such as bamboo sticks, mats and pulp. In the middle of the 1990s, Vietnam started to import bamboo processing machines from China, and started industrial development of the bamboo sector. According to statistics from the Anji Bamboo Machinery Society up to 2010, Vietnam had already purchased more than 100 production lines from Anji. On the basis of the successful

development of bamboo stick processing, the processing of other products – mainly chopsticks, toothpicks and incense sticks soon developed, such as bamboo pressed lumber, bamboo engineered boards, bamboo shoots, bamboo charcoal, bamboo constructions and furniture, etc. Many featured and beautiful handicrafts were also scaled up in production. In this way, the whole Vietnamese bamboo sector became comparatively large scaled and comprehensive, and its production value in 2010 reached US$250 million.

The cases of Anji, Lin'an, Xingyi and Vietnam introduced above show that they all had their own breakthrough points in their periods of development, and they all achieved success. This was because the breakthrough points were suitable for their own particular conditions. There are many other such cases, but there have also been failures or not very successful cases, such as the case in Ecuador described in Chapter 3 (Section 3.1.3) or the bamboo floor

Fig. 4.2. Bamboo charcoal honeycomb briquettes are used commonly in Ethiopia for the traditional coffee serving ceremony (Fu Jinhe).

Fig. 4.3a,b. Bamboo charcoal in the local market in Ethiopia (Fu Jinhe).

company in Vietnam discussed in Chapter 3 (Section 3.6.4, Case study 7) and in Annex 2. The reason of the failure is mainly because the lack of a thorough investigation and evaluation of the local conditions. To sum up, good decision making on the breakthrough point is important for success.

4.3   The Status of Bamboo Resources Is the Fundamental Basis of the Bamboo Sector

We have already discussed the problem of bamboo resources in Chapter 3, Section 3.1, but here I would like to stress this issue from a different point of view.

4.3.1   The utilization of natural bamboo resources

Natural bamboo forests without any rehabilitation and management are not suitable for industrial utilization, and presently, many countries and regions have large areas of natural bamboo forests that have not had any management activities. In some cases, the forests may have been planted, but left without management for a long period of time. For example, in Africa, the most widely distributed species is *B. vulgaris*, a species

Fig. 4.4. Bamboo firewood in the local market in Ethiopia (Fu Jinhe).

Fig. 4.5. Culm sections from *Schizostachyum funghomii*, which has long internodes (80–120 cm). It is very suitable for weaving products. (Zhu Zhaohua).

that was introduced from Asia a long time ago, but nobody knows when and by whom, and is now growing in Africa without any scientific management, basically in the wild. Natural forests of indigenous species and introduced species may have distinct disadvantages in the following aspects:

- they are a mix of different species;
- they are of unclear ages;

Fig. 4.6. *Dendrocalamus giganteus* in Yunnan Province, China (Yang Yuming and Hui Chaomao).

- they have a varied structure of culm ages in the same clump;
- they are comparatively seriously attacked by pests and diseases;
- they are of low quality;
- the stems in a clump are too crowded, which is not convenient for harvesting, or good for the growth of new stems;
- they are usually sparsely distributed, and so hard to access, harvest and transport.

Therefore, scaled, industrial processing using such resources would be extremely hard.

Before we start to develop the bamboo industry, these natural forests need to be rehabilitated and managed for the purpose of industrial use.

In fact, the technology of natural bamboo forest rehabilitation is very simple and effective, and usually the above problems can be tackled and resolved in 2–3 years, when the production and the quality will both be improved. The rehabilitated natural forests will look similar to monopodial bamboo plantations, and we could call them managed natural forests. For practical operations in natural bamboo forest rehabilitation, the following principles need to be mastered:

1. Remove old stems, leave young stems; remove stems that have already withered as well as those above 5 years old, but maintain an adequate proportion of stems of different ages.
2. Remove stems that are in the middle of the clump, but keep the stems that surround the clump.
3. Remove weak, diseased and insect-attacked stems, and keep strong ones.
4. Remove shoots generated from comparatively shallow rhizomes, but keep those germinated from deeper rhizomes for future mother stocks.
5. Remove condensed clumps but keep sparse clumps.

Here I would also like to stress that before investment, it is very necessary to study carefully and know how many resources in the local area are suitable for the intended product, and how many resources could be industrially processed. These figures should be actual figures after real investigation by experts, rather than figures for the total bamboo forest area provided by the local forestry sector; such figures usually seem very appealing, but they may not be able to tell you about the resources you can use for industrial purposes.

4.3.2 Develop plantations oriented towards a purpose – the principal pathway to a healthy bamboo sector

To make the bamboo sector a pillar sector in the local area – one that has a significant impact on the development of the local economy, on ecological construction and poverty alleviation, besides the protection, rehabilitation and management of the existing natural forests, it is necessary to develop plantations that are oriented towards a particular purpose.

To raise a plantation oriented towards a particular purpose, it is, of course, necessary to identify the purpose(s) of the plantation – whether it should be planted mainly for bamboo culm processing, for shoot processing, for handicrafts or for landscaping or urban greening, for reclamation of mined areas, or for water and soil conservation, etc. Oriented cultivation is to clarify the major functions and uses of the plantations being raised, but this does not mean that the plantation may not be able to play multiple functions in the future. For example, a plantation for culm processing purposes may also produce bamboo shoots and other products, and at the same time, may play a role in water and soil conservation. A bamboo plantation without a planned scale,

Fig. 4.7. Bamboo plantlets from nurseries being transplanted to the field in Ethiopia (Fu Jinhe).

quality control and certain management technology standards, would not be able to support a successful bamboo sector that could have wide and great impacts. Therefore, we hope those investors that are interested in bamboo, should put more of their zest and energy into investment into the development of bamboo resources.

Plantations also need to be developed according to the resource status and long-term development plans of different countries or regions. For example, presently, the mainland of Africa lacks large-sized bamboo species, fine shoot producing species and ornamental species for urban landscaping; this lack has limited the development of bamboo panels, special construction materials and bamboo shoot products, and even the bamboo gardening and landscape development. Hence, it is necessary to introduce and trial various fine bamboo species, and develop oriented plantations according to the strategic planning of the country; these actions are essential for bamboo development in Africa.

Bamboo is one of the easiest plants for sustainable management, as it has strong regeneration capacity, and once planted, takes only 3–4 years to generate income. Bamboo plantations need low investment, but have a high rate of return, and will also generate long-term, distinctive and positive impacts on the local economy, ecology and society. They are also a stable base for sustaining the bamboo sector.

4.3.3 Making accurate assessments of the potential utilization of different bamboo species

In Chapter 3, Section 3.1.2, we mentioned the utilization of different bamboo species, and stressed that a utilization strategy needs to be identified for each of the major species. In actual practice, though, people are not usually able to make accurate assumptions on the potential uses of a bamboo species. They either underestimate

Fig. 4.8. The culms in this clump of the bamboo *Bambusa vulgaris* look straight from far away (Zhu Zhaohua).

Fig. 4.9. When they are looked at closely, these culms of the bamboo *Bambusa vulgaris* are not straight (Zhu Zhaohua).

or overestimate, and these inaccurate assessments of the potential uses of bamboos may result in failure of the investment(s).

For example, some African countries have started to pay attention to the development of their bamboo sector, and some international organizations would also like to provide support, which is good. However, in Africa, there are only a few bamboo plantation projects, and most of the bamboo development projects have to rely on the existing resources. There are four main bamboo species in mainland Africa. One is *B. vulgaris*, as mentioned above; this bamboo usually grows below 1500 m altitude. The other three species are *Arundinaria alpina* (syn. *Yushania alpina*), *Oxytenanthera abyssinica* and *Oreobambos buchwaldii*, and they are distributed at different altitudes and are indigenous in Africa.

For industrial processing purposes, these four different African species have their own specialties. For example, *B. vulgaris* grows fast, has high biomass, wide distribution, high adaptability, is middle sized and can be easily propagated and planted. This species could be of comparatively large advantage if it is utilized for the following products:

1. Bamboo energy products – biomass energy pellets for power generation and charcoal. *B. vulgaris* is a good species for developing energy plantations.

2. Bamboo fibre products: pulp and papermaking, textile products and industrial fibre products, as well as fibreboards and particle boards; Brazil has a large area of plantations for pulp purposes that use this species.
3. Construction and furniture materials: scaffolding, structures, bamboo pole furniture, processed bamboo poles, etc.
4. Engineered products: pressed bamboo material and the series of products made from this, including bamboo sticks (toothpicks and barbecue skewers).
5. Bamboo handicrafts (including bamboo carving).
6. Ecology projects: for water and soil conservation, *B. vulgaris* can bring effects fast; it can also be used for small watershed conservation, riverbank and lake bank protection, the recultivation of mined lands, etc.
7. Urban greening: *B. vulgaris* is a cultivated variety, and *B. vulgaris* cv. Vittata is a famous urban greening material in Africa that has already been widely utilized; if it is mixed with *B. vulgaris*, there may be even better ornamental effects.

However, *B. vulgaris* has certain disadvantages: its splitting properties are not good, so it is not suitable for high value-added and fine bamboo woven products; the culms are not straight, so it is not suitable for sticks or strips longer than

Fig. 4.10. A bamboo fence made using the bamboo *Arundinaria alpina* (Zhu Zhaohua).

100 cm, and when processed into long strips or sticks, the utilization rate would be quite low. So *B. vulgaris* is not good for producing mats, curtains and laminated boards or cement moulding boards. Also, because it is not a large-sized bamboo, it is limited in use for producing some products that need larger sized bamboo species; neither is it good for shoot purposes, as described by Zhu Zhaohua and Li Dezhu in a report from a trip to Accra, Ghana, in 2002.

A. *alpina* is a species that is distributed at higher altitudes, with most of the natural stands concentrated between 2290 and 3360 m altitude. It may be one of the long-necked pachymorph/sympodial rhizome species, but the forests of *A. alpina* look quite like a monopodial forest, and are comparatively easy to manage. This species is distributed widely in Africa, and can be seen in many countries, such as Tanzania, Zambia, Kenya, Uganda, Rwanda, Zaire, the Congo (Democratic Republic of the Congo and Republic of the Congo), South Sudan, Ethiopia, Malawi and Cameroon, etc. It has a number of advantages in industrial utilization. For example, the forest distribution is concentrated, and the forests are usually large

and connected with each other, which is convenient for harvesting and management. As a middle–small sized bamboo, *A. alpina* has straight stems and good splitting properties. It is suitable for making bamboo woven products, furniture and constructions. The fibre content of the species (as dried bamboo) is on average 47.5%, and biomass production is also very high. If it is clean-cut, the average production capacity/ha can reach 100 tons (dried bamboo). Therefore, it is very suitable for energy or fibre plantations. The thicker bamboos of the species could also be used for bamboo mat/curtain production.

O. abyssinica is also widely distributed in Africa at altitudes between 1100 and 2100 m. The main countries of distribution include Malawi, Zambia, Zimbabwe, Burundi, etc. The stem of this species is almost solid, and the height of the bamboo is 8–16 m. *O. abyssinica* has the highest density of the African species, so it is the hardest one. The species is highly tolerant to drought, and the shoots are edible and in Tanzania are used for the production of fermented wine (see Chapter 2, Section 2.2.4). *O. abyssinica* is highly valuable for developing charcoal and

Fig. 4.11. A bamboo forest of *Arundinaria alpina* in Ethiopia (Jayaraman Durai).

energy plantations. It is also suitable for furniture and construction.

From the above accounts of different species, it can be seen that their different features and characteristics lead to their respective development potentials.

The authors' understanding of bamboos in Africa are quite primary, the above opinions are for reference purposes only.

4.4 Pay Attention to the Development Planning of Enterprises Before Investing

4.4.1 A feasibility study before investing

We have already discussed in detail the necessity of regional development planning – the object and the content, the multi-participation, the implementation, the adjustments and the evaluation (see Chapter 3, Section 3.3) and do not need to repeat this. Here, we will discuss enterprise planning. The failed enterprises mentioned previously did not make a comparatively complete plan based on a thorough feasibility study and planning before investment. In such a case, a series of difficulties and obstacles often appear after investments have been made, and once they have been found, and are not found possible to overcome in a short period of time, failure will be unavoidable. A problem that investors might often encounter is that, when they decide to invest, the local government may not have paid enough attention to the bamboo sector, hence there is no planning. Under these conditions, the enterprise becomes a pioneer, and any success that it achieves will push forward bamboo sector development in the local area. In this case, the investors need to have thoroughly conducted a feasibility study and planning and, at the same time, have striven to win understanding and support from local communities and the government. Simple impulsiveness, solo playing and blind investment without thorough study must certainly be avoided. This is because the bamboo sector is very complicated, encompasses many fields and is closely related with groups of people of different interests.

4.4.2   The guiding principles of enterprise planning

While formulating an enterprise plan, there should be a few guiding principles. In addition to profitability, there are a number of other principles, and these include:

- meeting the interests of the various stakeholders;
- synchronized, balanced and sustainable management and protection of the local bamboo resources along with the development of the sector; and
- the promotion and conservation of the local ecological system and environment.

Predatory or extensive exploitation for short-term benefits must be avoided at all costs.

4.4.3   The main contents of enterprise planning

For bamboo sector development planning, besides the local plan guided by the local government, enterprise investment plans should also be included. The enterprise investment plan should include the following components.

Local investment conditions and resource assessment

1. A very accurate assessment of the quality and quantity of local bamboo resources is needed, including estimation of potential suitable products, and the amount of the resource that can be immediately utilized for industrial processing.
2. A detailed study of the local investment environment. How much attention does the local government pay to the bamboo industry? What is the local government's positioning of the bamboo sector? Which part of the existing policy system is positive? Are there negative policies and, if so, how should these problems be solved?
3. What are the local traditional bamboo products? What is the local community culture, and what is the personnel capacity?
4. What are the bamboo related government sectors and non-governmental organizations.
5. If there are enough and stable power and water supplies, etc.

Development goals

1. What are the main investment orientations, e.g. bamboo resource cultivation, bamboo culm processing, shoot processing, handicrafts or ecological and cultural industries, etc.?
2. What are the production scales, main products and product structure? Define the main products and the other associated products and series and, at the same time, provide a clear goal of the scale of production.
3. What are the expected benefits? These should include the input and output benefits of the enterprises, the estimated driving power for local income and employment generation, and the benefits for local environmental improvement.

Implementation

Objectives, missions and indicators need to be identified for all activities in the plan.

Specific measures guarantee mission completion

The conditions for and the practical measures that must be taken to ensure the success of the plan include:

- provision of the necessary lands and funds;
- local government policies towards allowing the plan;
- the cooperative mechanism of the enterprises with local governments and communities;
- training of the employees and partners;
- satisfaction of technical codes and product standards;
- establishment and completion of the supply chain (from raw materials to processing and to markets); and
- the necessary research projects and technical innovation mechanisms and their specific targets.

The associated budget

The budget associated with the enterprise plan needs to be included in the plan itself.

Estimation of impacts after implementation of the plan

These include:

- impacts on the enterprise itself – product quality and quantity; meeting the demands of the domestic and international markets; input–output analysis; etc.; as well as
- impacts on the local economy, ecology and society, including benefits to the local social economy, sector development and community income, increased employment opportunities, impacts on local environmental conservation, e.g. carbon sinks and storage, and so on.

4.4.4 A good plan is a crucial part of the way towards success

A good plan will provide an enterprise with clear goals for future actions, and specific methods and steps, practical measures and the necessary funds to achieve the goals. These will help the enterprise to always keep a sober mind, initiative and a clear understanding of the situation. Similarly, it will allow all of the stakeholders to fully understand the content and schedule of the enterprise, so that they can provide the necessary support and active cooperation. A good plan is also conducive to obtaining support from the relevant government and international organizations.

Of course, good planning is not to put things right once and for all. Although assumptions have been made on the various challenges and difficulties the enterprise might encounter, and measures have been identified to overcome them, even the best plan may not be able to avoid the unpredictable. The unpredictable may be positive, such as more support from investors or funds, but it may also be negative, such as technical bottlenecks, etc. Consequently, the necessary adjustments to the plans are allowed when they are essential. However, this point should not neglect the necessity and importance of planning, and it can be stated without any exaggeration that a good plan provides a solid base for the success of an enterprise. A good plan depends on the thorough investigation of local resources and social conditions, and a degree of understanding of the current world bamboo development situation; it also depends on a clear awareness of the strengths and weaknesses of the enterprise. According to the principle of 'playing the competitive edges to the full and avoiding the weaknesses', an enterprise may identify a realistic, practical and feasible plan that could help it meet its goals through continuous efforts, but is not prone to extravagant fancies, is not unrealistic and does not follow a stereotypic routine that lacks creativeness.

4.5 The Government Should Play an Active Role in Bamboo Sector Development

We have discussed the various factors affecting the development of China's bamboo sector (see Chapter 3, Section 3.2). The roles of the government are indispensable to all these factors and the government is crucial to the development of the bamboo sector.

4.5.1 Why should the roles of the government be emphasized?

The bamboo sector encompasses a wide range of different fields, and is closely related to many social groups with different interests. While developing the local bamboo sector, all stakeholders need to cooperate with each other: the farmers, processing enterprises, grass roots level administrative sectors, community organizations, marketing practitioners, and so on. The development environment of the bamboo sector is always more complicated than that of other industrial sectors and high-tech sectors. Another feature of bamboo sector development is that its success or failure always brings direct and immediate impacts on the local poor communities – income, employment, rural development and environmental improvement. Because of the above features, the government's involvement in the development of the bamboo sector is indispensable. Without the support of and coordination by the government, an enterprise may find it is extremely hard to develop well, as it planned, and many key problems may not be solved in an appropriate way.

4.5.2 Identifying the local bamboo sector development plan should be prioritized in the government's agenda

In order to support the local bamboo sector, a government should first identify a plan. A good local plan indicates the degree of attention attached to the sector and the positioning of the sector by the local government. It will also include the government's goals for development of the sector, and the strength of its support, measures and policies for implementation. A local plan will encourage and strengthen the confidence of the investors, and allow better contribution of the investors to the consummation of local sector planning.

4.5.3 The government should make full use of the leverage roles of its policies to promote the healthy development of the bamboo sector

A very important role of the government is to identify a series of preferential policies for the bamboo sector, based on a thorough study of the needs of stakeholders and a good understanding of the features of the bamboo sector. The bamboo sector is related to policies on various other aspects of government responsibility, and policy making for the bamboo sector needs to prioritize the most urgent issues that need resolution according to the local conditions. Let us take Anji and Lin'an – the best development models from China – as examples.

Before the reform and opening up policy in China, the land in the rural areas, including forest land with bamboo forests, was largely managed by collectives. In contrast, after the reform and opening up, local governments prioritized the resolution of the land user rights and forest management rights, and allowed farmers to have self-managed bamboo forests or have user rights of the land to plant bamboo. This policy greatly encouraged the farmers to participate in bamboo resource development. Another policy change also greatly encouraged

the local bamboo farmers to establish microenterprises, and introduced foreign investment to the local bamboo sector.

After these policies had been implemented, the Anji and Lin'an governments invariably focused on the construction of their demonstration bases. The construction of successful demonstration bases could provide a solid base for the future development of the bamboo sector. Successful cases would be able to increase the confidence of the investors and a series of other related government departments, and encourage more stakeholders to join the sector. The construction of these demonstration bases included various approaches. In Anji and Lin'an, a series of demonstration bases were established at different development stages of the bamboo sector. These included: high-yielding bamboo plantations and demonstration households, villages; non-polluting plantations for shoot purposes; low-yielding bamboo forest rehabilitation; plantations for both shoot and culm purposes, and plantations for sole shoot purposes; trials on transferring collectively owned enterprises to private ownership; trials on free interest loans for bamboo shoot processing enterprises; bamboo-featured ecotourism and farm staying; professional bamboo shoot markets, bamboo product business streets and trade centres, etc. The emergence of a series of successful demonstration bases allowed the Anji and Lin'an bamboo sectors to develop fast and under stable conditions for a long period of time. A group of professional personnel was cultivated during the construction of the demonstration bases, and has driven the power for more partners to actively participate in bamboo development.

In addition to the construction of the demonstration bases, Anji has developed other series of policies at different development stages of the bamboo sector. For example, its tax policies, farm-stay policies, investment policies, research and innovation policies, etc. When identifying policies for bamboo sector development, all countries and regions need to first focus on the key problems that hinder the development processes of that sector, and identify corresponding policies to solve the problems.

For example, I used to visit two countries in West Africa – Ghana and Nigeria. In those two countries, most of the lands belong to the Chiefs (or Kings) of the local tribes (communities).

Tribe members could apply to the Chief to be allowed a certain area of land to plant crops. However, the ownership of the bamboos was not clear, and there was also no bamboo management, although community members could harvest bamboo freely and according to their needs. This situation was not conductive to the active participation of the community in the cultivation and management of bamboo. It also did not help the efforts of enterprises to establish and develop businesses, as without stable, sufficient and suitable supplies of bamboo raw materials, the enterprises were not able to actively invest in the bamboo sector. Neither did the situation help with cooperation between enterprises and local communities in bamboo resource cultivation.

Under such a situation, how to coordinate the relationship and benefits among the enterprises, the Chief and the community members, and how to encourage their cooperation and active participation in the bamboo sector, are the key issues to be resolved. Our suggestions are: that the Chief could allow certain areas of lands for households who are interested in bamboo management, or allow those whose leased lands already have bamboo resources to access the management rights of the bamboo forests and, at the same time, have ownership and disposal rights of the products. The enterprises may provide bamboo plantlets and technical training to the local communities, and agree with the communities to purchase the bamboo raw materials according to market prices. The enterprises and the community may negotiate with the Chief, and reach an agreement on the lease rate of the lands; the Chief could also play the role of coordinating the relationship between enterprises and community members. Besides cooperating with community households in managing and cultivating the local bamboo resources, the enterprises may also directly rent lands from the Chief. This joining of enterprises in bamboo resources management will also bring more employment opportunities and will also generate revenue for the government.

Such a cooperative model could be called: 'enterprise + Chief + community households', and it is a triple-win model. On the one hand, the three parties in the model are interdependent, but on the other hand, they are relatively independent; yet they are also partners that have integrated interests. The local government should continue to follow the operation of the model, and coordinate and strengthen the partnerships of the three parties. If conditions allow, certain funds or technical support should be provided; and these should be very much welcomed by the stakeholders. The cooperative model introduced above should be stabilized through mutual understanding and trust among the three parties; what is more, it should be consolidated through the signing of legal agreements witnessed and supervised by the local government.

To sum up, the policy of the government provides important leverage to the development of the bamboo sector. Good policies may attract investors, and inspire initiatives by people from the community, so, we could say that sometimes the correct policies are more important than money.

In support of the development of bamboo sector, the government should gradually develop a comprehensive policy system. The aim is to create a favourable environment, and promote the sector's sustainability. This system should include:

- fully mobilizing the enthusiasm of the enterprises and people in the community to participate, and allowing them to access the benefits due;
- facilitating close cooperation among stakeholders, creating a mutual benefiting and multi-win situation;
- protecting, enlarging and stabilizing the local bamboo resources, aiming at high quality and sustainable management;
- continuously prioritizing and upgrading the industry structure;
- encouraging product and technology innovation; and
- improving the local environment and ecology, etc.

4.5.4 The government should provide appropriate financial support when necessary

Because the bamboo sector is closely related to public benefit missions, such as community development, poverty alleviation and ecosystem construction, beside identifying preferential policies, the government should also provide certain financial support from different channels. Direct financial support from the government should be 'seed' money, i.e. using small funds to generate large impacts. For example, providing:

- free plantlets to households that find it difficult to manage;
- partial loans to enterprises;
- technical training to community households;
- expert information to enterprises;
- project funds for research and technical transformation;
- prizes and awards for outstanding contributors to the bamboo sector;
- appropriate reduction of or exemption from taxes and fees for pioneer investors; and
- overseas study tours and international exchange opportunities for key investors and partners of the bamboo sector, etc.

Another type of financial support is bamboo development projects, for which the support could be direct or indirect. For example, the construction of forest roads to facilitate the transportation of bamboo raw materials and products could be included in the infrastructure construction plans of the government. Another example would include bamboos in local ecological construction and restoration projects, such as water and soil conservation, micro-watershed protection, rehabilitation of mined areas, landscape construction, rehabilitation of degraded lands, etc.

Contracts can be reached with investors and local communities to confirm cooperation. These should be allowed for bamboo plantation and management, and will not only generate ecological effects, but also solve the raw material supply. Bamboo projects may not only help the government to meet its goals and requirements in ecological construction, but also help local communities and enterprises to save on investment, thus achieving a multi-win situation. According to the needs for development of the bamboo sector, the government could join efforts with the enterprises in allowing certain lands and finance for establishing 'Bamboo Industry Zones', 'Enterprise Zones' or a 'Bamboo Ecology, Trade and Culture Park'. These will help to attract both investors and technical cooperation

from universities and research institutes, promote trade and provide platforms for cultural tourism.

There are many more potential ways and channels of government support for the bamboo sector that could be explored. Special note needs to be taken of public benefit projects and social development projects in which bamboo could play important roles. In Chapter 3 (Section 3.6.3, Case study 6), we introduced the case of Sichuan Qingshen Yun Hua Bamboo Culture Travel Ltd. in Qingshen County, Sichuan Province; this company has been providing training for local poor rural households, women, disabled people and minority ethnic people on bamboo weaving techniques since 1984, under long-term support from the local government.

4.5.5   Provide perfect services and avoid blind decisions

We have discussed the important roles of the government in bamboo sector development, as well as how or in which aspects the government should provide support and services. It is also necessary to see what the government should be cautious not to do.

When there are things that should be determined by enterprises and community households by themselves, the government should not interfere, take over or replace these roles. In China, the local governments of a few bamboo producing areas have sometimes made such mistakes. For example, a government leader preferred one or a number of bamboo species, and without consulting the relevant experts and local enterprises and communities, made a decision to develop these bamboo species, and the enterprises and communities were called upon to participate in planting. Usually, such cases will fail because the bamboo species cannot be able to adapt to the local climate or soil conditions, or will not be accepted by the markets.

A similar situation may happen when wrong decisions are made on product orientation and structure. Which product or series of products should be developed should be decided by the enterprises themselves, without government interference. If the enterprises need technical support, the government may invite appropriate experts to assist. The government should also not interfere with the international administration of the enterprises, but may provide services when needed. It should be clarified that the government's role is providing services, rather than taking over and making decisions for enterprises.

To sum up, in the course of bamboo sector development, the government should play active roles in guiding, supporting and coordinating. According to the needs of the various stakeholders, and the requirements of the development of the sector, the government should provide guidance for the demonstration of new development orientations, new technologies, new products and the introduction of new bamboo species. The government should also coordinate the partnership among stakeholders and partners directly or indirectly through the services of professional societies and non-governmental organizations (NGOs), so as to achieve win–win or multi-win purposes. In addition, the government should aim to provide a good investment environment for the bamboo sector, and guarantee its stable and sustainable development.

Annex 1

Abstract of the Development Plan for China's Bamboo Industries (2013–2020)

Being critical ecological, industrial and cultural resources of the world and commonly acknowledged as the 'second largest forests of the planet', bamboos make up an essential ingredient of the global forestry resources. China, one of the primary bamboo growers and known to the world as the 'Kingdom of Bamboos', ranks top of the world in terms of bamboo germplasm, sizes of bamboo groves, volumes of bamboo stock, and output and export volumes of bamboo products. Over the years, China has gone through speedy growth in both the sizes of its bamboo groves and its standing bamboo stocks. China also leads the world in product innovation as well as in R&D on bamboo processing techniques, and steadily supplies the world with thousands of bamboo products that fall roughly into dozens of categories, including, but not limited to, construction materials, household bamboo products, bamboo-based panels, bamboo charcoal, bamboo-based furniture, bamboo fibre and bamboo-derived drinks. In the year 2012, the total outputs generated from the bamboo industries in China, which provide jobs for over 7.75 million people, reached a value of US$19.5 billion. The bamboo industries have emerged as a dynamic force in China's forestry sector and offer tremendous potential for its overall economic growth and for enhancing household income in the country's rural areas. To guide and regulate the country's bamboo industries for their healthy development, the

China State Forestry Administration (SFA) compiled a Development Plan for China's Bamboo Industries (2013–2020). This is outlined in Annex 1.

A1.1 Geographical Distribution and Characteristics of Bamboo Resources in China

Over 500 bamboo species in 39 genera grow in China, including groves of tropical sympodial bamboos, individually growing subtropical monopodial bamboos and cold-resistant compound species that typically grow in high-altitude, high-latitude regions. Natural bamboo forests can be found in 27 of China's provinces, the exceptions being Xinjiang, Inner Mongolia and Heilongjiang, and are most densely distributed in the mountainous regions of 15 provinces – Fujian, Zhejiang, Jiangxi, Hunan, Guangdong, Hainan, Anhui, Hubei and Henan in the eastern part of the country, and Sichuan, Guangxi, Guizhou, Chongqing, Yunnan and Shaanxi in the western part of the country. Due to the sharply varied climate, soil and topological features of each region, and the biological distinctions between bamboo species, obvious regional characteristics can be identified in the geographical distribution of bamboos in China. The four primary regions are as follows:

1. The Yellow River–Yangtze River Region. This is situated between 30 and 40° in the northern latitude, where the annual average temperature falls between 12 and 17°C, and the annual rainfall ranges between 600 and 1200 mm. The predominant genera/species include: *Phyllostachys sulphurea* cv. [var.] *viridis* (syn. *Phyllostachys viridis*), *Pleioblastus amarus, Fargesia, Arundinaria, Sasa longiligulata, Bashania fargesii*, etc.

2. The Yangtze River–Nanling Region. This is situated between 25 and 30° in the northern latitude, where the annual average temperature falls between 15 and 20°C, and the annual rainfall ranges between 1200 and 2000 mm. Being the largest bamboo-growing region in China, this region is home to such predominant species as: *Phyllostachys sulphurea* cv. [var.] *viridis* (syn. *Phyllostachys viridis*), *Pleioblastus amarus, Brachystachyum densiflorum, Indosasa angustata, Neosinocalamus affinis, Chimonobambusa quadrangularis*, etc.

3. The South China Region. This is situated between 10 and 20° in the northern latitude, where the annual average temperature falls between 20 and 22°C, and the annual rainfall ranges between 1200 and 2000 mm. This region has the richest diversity in bamboo species, with such predominant genera/species such as: *Bambusa, Dendrocalamus, Acidosasa chinensis, Dinochloa, Gigantochloa, Bambusa cerosissima, Pseudosasa amabilis, Pseudostachyum polymorphum, Schizostachyum chinense, Melocanna baccifera*, etc.

4. The South-western High-altitude Region. This is the 1000–3000 m high mountainous region in South-western China, where the annual average temperature falls between 8 and 12°C and the annual rainfall ranges between 800 and 1000 mm. This region has the most densely concentrated primitive bamboo groves, with such predominant genera/species as: *Chimonobambusa quadrangularis, Fargesia, Qiongzhuea tumidissinoda, Yushania, Neosinocalamus affinis*, etc.

By the end of 2011, the area of bamboo forests in China had reached 6.7274 million ha, with 28.622 billion standing bamboo plants. This can be broken down into the following categories: 373,800 ha of shoot-producing bamboo forests (5.56% of the total); 940,600 ha of pulp-oriented bamboo forests (13.98% of the total); 2,443,600 ha of timber-producing bamboo forests (36.32% of the total); 1,621,700 ha of dual-purpose forests (i.e. those that produce both bamboo shoots and timber; 24.11% of the total); 1,255,700 ha of eco-oriented bamboo forests (18.67% of the total); and 50,500 ha of landscape-building bamboo forests (0.75% of the total). In addition to this, the accumulated sizes of bamboo nurseries were measured at 3800 ha, with 120 million individual bamboo plants turned out annually. Forty-three parks (with a total size of 35,000 ha) and 26 botanic gardens (with a total size of 969 ha) that feature bamboos have been established throughout the country, together with an additional 181 ha of miniature bamboos that are specially cultivated for creating potted landscapes.

A1.2   Current Development of the Bamboo Industry in China

A1.2.1   Current situation of bamboo resources

The area of bamboo forest expanded sharply as a result of the active policy to develop bamboo resources, and the total area of bamboo increased from 2 million ha in the early 1950s to 6.73 million ha in 2011. In the meantime, the operation and management of bamboo forests made great progress by the use of intensive managerial measures that included the transformation of the low-yielding forests and artificial cultivation. The overall quality of the bamboo forests was also greatly promoted, so that the average number of standing bamboo plants per mu (1 ha = 15 mu) increased from less than 90 in the 1970s to 138 currently, and the average diameter at breast height (DBH) increased from 6 to above 8 cm. The species structure of bamboo has also been perfected through the development of market-friendly and high-value bamboos, such as *Dendrocalamus giganteus, Phyllostachys praecox, Dendrocalamus latiflorus, Dendrocalamopsis oldhami, Acidosasa edulis* and *Pleioblastus amarus*, besides the cultivation of *Phyllostachys*.

A1.2.2   Current development of bamboo shoot processing

Two hundred species of bamboos in China are edible among the total 500, and 30 of these are of high quality, including *Phyllostachys praecox, Dendrocalamus latiflorus, Dendrocalamopsis oldhami* and *Acidosasa edulis*. Most of the fresh shoots are cooked instantly, while still some shoots could be processed into varieties of packaged snack products. It is estimated in 2011 that the total annual production of bamboo shoots is 1.66 million tons, mainly exported to Japan and South Korea.

A1.2.3   Current development of bamboo-based panels

For 20 years, the bamboo-based panel industrial system has been formulated based on the abundant bamboo resources in the country and the experience learned from wood-based panel production. There are thousands of bamboo-based panel processing enterprises nationwide, which are mainly located in provinces rich in bamboo resources, such as Zhejiang, Hunan, Sichuan, Jiangxi, Anhui and Fujian. There are also dozens of categories of products, such as bamboo woven plywood, bamboo plywood, bamboo laminate, bamboo pulp plywood, bamboo fibre panel and bamboo shaving board. The total bamboo-based panel production in the year 2011 reached 2.5 million tons.

A1.2.4   Current development of bamboo flooring

Bamboo flooring is a self-developed product in China, and with its characteristics of hardness, flexibility and steadiness, it has won the love of consumers since its appearance in the domestic market in the 1980s. More than thousands of processing enterprises around the nation ensure the leading position of bamboo flooring in both its processing technology and product quality. The total production of bamboo flooring in 2011 reached 779.6 thousand m³ and about 60% of this production has been exported to more than 40 countries and regions, making China the primary production and export base for bamboo flooring worldwide.

A1.2.5   Current development of bamboo pulp and papermaking

Bamboo pulp can either be 100% bamboo or a mixture of bamboo pulp, wood pulp and grass pulp in a certain ratio. Both the pure and the mixed pulps can be processed through the procedures of papermaking. China has the advantage of low-cost bamboo due to its rich bamboo resources, and the production of bamboo pulp in 2011 was about 1.52 million tons.

A1.2.6   Current development of bamboo utensils (handicrafts)

With their long history and variety, Chinese bamboo products cover a huge range in daily life from interior decoration to gifts, packages, tableware, kitchen utensils, garden tools and yard construction materials, so they have potential market prospects. China is a pioneer in scientific research on and the development and utilization of bamboo utensils and the production in 2011 was about 2.63 million tons.

A1.2.7   Current development of bamboo furniture

Bamboo furniture refers to furniture composed of bamboo, bamboo panels and/or bamboo boards. Recently, the utilization of bamboo panels and boards has become widespread due to the perfection of their processing technology.

A1.2.8   Current development of bamboo fibre products

The bamboo fibre products industry in China has formed a production chain with segments of bamboo pulping, spinning, weaving, dyeing and processing at a relatively fast speed. In 2011, there were 59 enterprises in this field producing various bamboo fibre products of high quality, which have won acceptance from consumers; total production in that year was 96.3 thousand tons.

A1.2.9   Current development of bamboo charcoal processing

While bamboo charcoal products appeared in China 2000 years ago, the modern industry was actually developed in Japan 20 years ago.

In recent years, Chinese processing technology, skill and facilities have surpassed those in Japan and reached the leading position in the world, so that even Japan needs to import about 4 thousand tons of bamboo charcoal from China annually. At present, the products cover ten series over 100 categories, and include bamboo shavings charcoal, bamboo trunk charcoal, utensils in daily use, humidity regulators, deodorants, cleaning detergents, handicrafts, health supplements, purifying materials and shielding material. The total production of bamboo charcoal in 2011 was about 136.2 thousand tons.

A1.2.10   Current development of bamboo-derived drinks

Since the first bottle of bamboo juice was produced in 2004, bamboo-derived drinks, which are characterized by their advantageous functions in body regulation, such as diuresis, lowering the blood fat and so on, have acquired the acceptance of increasing numbers of customers. It is estimated that the production of bamboo-derived drinks in 2011 was up to 17.425 thousand tons.

A1.3   Developmental Plans for the Chinese Bamboo Industries

A1.3.1   Scope

Taking into consideration the geographical distribution of bamboos and China's natural conditions, this plan identifies 892 counties in 16 provinces (municipalities/autonomous regions) that are endowed with favourable natural conditions enabling them to act as key areas for developing bamboo industries. These include Fujian, Zhejiang, Jiangxi, Hunan, Sichuan, Guangdong, Anhui, Guangxi, Hubei, Guizhou, Chongqing, Yunnan, Jiangsu, Shaanxi, Henan and Hainan.

A1.3.2   Overall guidelines

The distribution of industries needs to be optimized by boosting the growth of advantageous bamboo species in regions with the most favourable natural conditions so as to reap the benefits of economies of scale, which will, in turn, contribute to the formation of integrated belts for bamboo industries that take the fullest advantages of their corresponding regional features and their most celebrated brands, as well as their most competitive enterprises. The management of bamboo resources is to be strengthened, with focus placed on upgrading the low-productivity bamboo forests that are currently in existence. The cultivation of new bamboo cultivars that are characterized as high quality, high output and high efficiency as a means of speeding up, will be carried out in combination with technology-innovation initiatives aiming at improving management practices and promoting eco-friendly notions in the development of new products. Efforts will be made to expand the scope within which bamboo fibres and bamboo-based panels can be applied, so as to widen the markets of bamboo-derived products and facilitate their export. C&I systems and quality supervisory systems will be put in place to ensure that the bamboo-growing and product-processing meet corresponding technical standards.

A1.3.3   Distribution

In line with the current situation and developmental potential of bamboo resources and the corresponding industries in China, and in terms of the natural and topological differentiation and distribution of bamboos, the Plan divides the bamboo-growing areas of China into two large categories, namely 'Prioritized Development Counties' and 'Standard Development Counties', as elaborated in detail below.

Prioritized Development Counties

This category covers a 14,000 ha area of bamboo forests in 150 counties. The vision for these counties centres on upgrading the quality and the production of the bamboo forests, with efforts to be made to develop modernized bamboo industrial parks that serve as incubators for large-scale enterprises; these, in turn, will contribute to the formation of horizontally and vertically integrated industrial chains where large-, medium- and small-scaled businesses can all develop healthily and in balance. In addition to sharpening the competitive edge of the bamboo industries, these regions are also expected to play pioneering roles in exploring efficient mechanisms and modes that will facilitate the booming of the bamboo industries.

Standard Development Counties

This category covers the rest of the 742 counties in the scope of the Plan, which are designated as standard development counties. The vision for these counties is that resource conservation will be the fundamental task, but that efforts will also be made to promote the technologies involved in bamboo resource intensive management, so as to facilitate the transfer from the current extensive management mode to an intensive-management mode. Managerial standards and the productivities of bamboo forests will be improved to guarantee the sustainability of bamboo resources, hence laying down a reliable basis for the sound development of bamboo industries.

A1.3.4   Objectives

By 2015, the total value of the output from bamboo forests in China is expected to reach US$32 billion, increasing by 66.5% from that in 2011. By 2020, the total value of the output of the bamboo forests will reach US$48 billion, with 10 million people directly involved in labour in the bamboo industries, and the income of rural households from bamboo industries averaged over the prioritized and standard development counties is expected to be US$330/person, accounting for 20% of the net income/person in rural areas.

A1.4   Key Tasks for China's Bamboo Industries

A1.4.1   Bamboo resource cultivation

The establishment of bamboo resource bases

Innovative management approaches, such as purpose-oriented cultivation, diversified management, etc., will be adopted to promote intensive

management in low-efficiency, low-productivity bamboo forests. Measures will be taken to protect the biodiversity of bamboo forests so as to enhance their corresponding ecological functions. It is planned that the total area of bamboo forests in China will be raised to 4 million ha, of which 1 million ha will be newly developed and the other 3 million ha will be upgraded from existing stands. Four key bases will be established: shoot-producing stands, pulp-oriented stands, timber-producing stands and dual-purpose stands for generating both bamboo shoots and timber.

The establishment of bamboo nursery bases

In line with development needs and existing (as well as potential) conditions, bamboo nursery bases, which will adopt different propagation methods (such as growing/breeding from seed; separating plant seedlings; raising young plants from cuttings or by layering or rhizome shooting; and mother bamboo planting) and will be set up in 13 provinces (municipalities/autonomous regions) covering 237 counties. Across the country, 411 bamboo nurseries are planned, covering an area of 5548 ha. In addition, 242 existing nurseries, with a total size of 3438 ha, will undergo renovation. The annual supply from these bamboo nurseries is expected to reach 500 million plants.

A1.4.2   Construction of the bamboo product processing industry

With due attention to resource endowments and the existing foundations of bamboo processing industries, regional characteristics will be highlighted in mapping the corresponding courses of industrial development in each region. While the development of less resource consuming but high-added-value bamboo industries such as ecotourism, bamboo fibre manufacturing, the production of bamboo-derived drinks and bamboo handicrafts, bamboo shoot processing and bamboo charcoal production will be prioritized, the more resource-consuming industries, including those engaged in making bamboo-based panels, bamboo flooring, bamboo-based furniture and bamboo-generated pulp, etc., should only be developed on moderate scales.

Bamboo shoot processing

Based on the current situation, efforts should be made on the one hand to perfect traditional shoot products, increasing their taste and added value, and on the other hand to research and develop new products and instant products with various flavours to meet the dietary habits and pace of living in different regions and countries, in order to enlarge the international markets.

Bamboo pulp and papermaking

Taking advantage of the favourable geographical location and resource endowments of the provinces in the coastal regions of eastern China and the south-western region, and bearing in mind the status quo of the pulp and papermaking business in terms of both geographical location and developmental trends, more attention should be paid to developing this industry according to the scale of the local economy.

Bamboo-based panels

Focus will be placed on adding variety to the existing bamboo-derived panel products and expanding their scope of application, including the development of integrated bamboo timbers, remoulded bamboo wood, bamboo plywood, bamboo-wood composite boards used as moulds for prefabricated concrete slates, bamboo-wood composite boards used in making shipping containers, bamboo furniture, bamboo flooring, and bamboo plywood used as car or train carriage floors.

Bamboo flooring

Measures will be taken to facilitate the integration of the bamboo flooring manufacturing industries and bamboo growing industries, gradually bringing into existence an industrial pattern that involves tri-stakeholders (business + growing bases + farmers). R&D will be strengthened to turn out more product varieties, and the intensive management of scaled businesses will be encouraged to create celebrated brand names in the industry.

Bamboo furniture

Centring on the key regions that have traditionally been known for the furniture-making industries, innovative measures will be adopted to apply advanced processing and design technologies to the development of new products, so that the varieties of bamboo furniture and their corresponding added values can be further increased. Marketing campaigns will also be launched extensively to improve the consumer awareness of such products and, in turn, boost their market share.

Bamboo fibre products

R&D programmes will be launched vigorously to update the processing techniques used for extracting bamboo fibres and other new fibres. Efforts will be made to expand the fields where bamboo fibre and the corresponding products can be extensively applied. Regulations will be made to set standards for bamboo fibre testing and to facilitate the emergence of renowned brand names in this sector.

Bamboo-derived drinks

By applying advanced techniques, juice and essential nutrients will be extracted from bamboo plants and bamboo leaves to develop a rich variety of bamboo-derived drinks, including carbonated beverages, fruit juices, bamboo teas, bottled beverages, vegetable protein drinks and other functional beverages.

Bamboo handicrafts

Productive bases will be vigorously developed to turn out a huge variety of bamboo-made products that are used in daily life, such as chopsticks, sticks, bed mattresses and (other) household appliances. Efforts will also be made to boost the development of bamboo handicrafts used for ornamental/recreational purposes, including woven bamboo items, bamboo sculptures and bamboo tea utensils.

Bamboo charcoal processing industry

Extensive R&D projects will be carried out to develop a series of products that tap into the multiple functions of bamboo charcoal and extend the corresponding product lines. In addition to its industrial uses and uses in daily life, the use of bamboo charcoal in health building will also be actively tapped. Incentive mechanisms will be put in place to encourage some large-scale businesses to emerge as the leading forces of the sector.

A1.4.3 Spreading the cultural awareness of bamboo

In order to strengthen the cultural awareness of bamboo among ordinary civilians, efforts are to be made in the construction of bamboo forest parks and tourist scenic spots, which can help to develop eco-cultural tourism and utilize the cultural elements of bamboo. Additionally, efforts will be made for the protection of cultural inscriptions, the construction of museums and natural venues, and the production of editions of books and audio materials to enhance the development and utilization of bamboo culture and boost integration between bamboo culture and the eco-economy.

Both the 'Bamboo Culture Festival' and the 'Bamboo Shoots Culture Festival' are held to deeply investigate and promote bamboo culture. Bamboo culture museums and bamboo crafts museums are expected to be set up. Original production signs are to be applied to bamboo and bamboo charcoal products, and these signs will be utilized as a medium to promote the construction of the bamboo cultural industry, boost international studies on bamboo culture and attract bamboo and bamboo charcoal merchandisers to China.

A1.4.4 Developing supportive systems for the bamboo industries

Supportive systems for science and technology

First, a state-wide strategic alliance of technological innovation for the bamboo industries will be set up to create a platform for bamboo specialists and researchers in the academic field to carry out innovative studies and develop new talents. Secondly, market-driven incentive systems, which are led by businesses and integrate businesses, schools and academia into closely tied chains, will be adopted to encourage the application of technological breakthroughs to productive activities. Thirdly, promotional activities will be launched to disseminate new practical technologies so that new plant cultivars, new techniques, new methods and new productive procedures can be widely employed in the bamboo industries.

Supportive systems for raw materials, transactional activities and markets

Well-functioning, specialized markets for bamboo shoots and other bamboo products will be established to speed up the circulation of the corresponding goods, covering the whole process and ranging from raw material production, product processing, display and sales through to the final conclusion of the transactions. Next, a modern market transaction system for the bamboo industries will be established, which will combine agency, delivery and operation of the chain into an integrated whole, so that marketing campaigns can be carried out efficiently and effectively. Last, in accordance with the needs of the market, specialized economic cooperatives, trade associations and professional brokers will be organized so that their corresponding functions in boosting technological advancement, technical standard formulation, trade promotion, market access, public services, etc. can be tapped.

Standard formulation and quality warranty systems

1. A relatively sophisticated standards system, which includes systematic criteria and indicators (C&Is) and the corresponding certification schemes for the bamboo industries, will be established to speed up international recognition of China's bamboo products and their corresponding standards.
2. The quality warranty system will be further streamlined to facilitate quality control, product tracing and supervision in China's bamboo industries, with competent testing services duly set up.
3. Comprehensive systems will be established to guide the operations of businesses engaged in this line. A licensing system and chain-of-custody system will be adopted to put under control the whole process involved in the growing of raw materials and the processing and sales of bamboo products, and to counter all activities that infringe upon the rights of concerned parties.

Information systems

1. A comprehensive information system that traces the economic data of the bamboo industries will be put in place to keep a pulse on the development of the industries, with such functions as forecasting, consistent monitoring of key products and key enterprises, dynamic analysis, real-time inventory and prewarning.
2. An e-commerce platform for the bamboo industries, including transactional websites and e-stores, will be established to promote the digitization of this sector.

Enhancing publicity campaigns for the development of bamboo industries

The government, trade associations and enterprises will work closely to launch extensive publicity campaigns about the bamboo industries and their products so as to set up positive images of the industry among consumers. Various media channels, including the Web, newspapers, TV and others, will be used to run welfare-oriented advertisements concerning the benefits provided by bamboo products. In addition, bamboo-themed shows and trade fairs will be organized at both provincial and national levels to improve people's awareness of the industry and its significant roles in ecological conservation, environmental protection and health building. Through such wide exposure, people will be supposed to have better understanding of the characteristics and advantages of bamboo products, hence promoting the (market) awareness and influence of bamboo products and formulating a certain industrial atmosphere for utilizing bamboo and loving bamboo products.

Annex 2

Report on a Chinese Consultancy for The Bamboo Company in Vietnam

This is a report from a consultancy undertaken by a team of experts from China for The Bamboo Company in Vietnam; it was first mentioned in Chapter 3, Section 3.6.4, Case study 7.

A2.1   Introduction

As invited by one of the bamboo companies in Vietnam, henceforward designated 'The Bamboo Company', a team of experts from China conducted a consultancy visit to the company's factories in a village. The expert team was composed of Prof. Zhu Zhaohua and a bamboo engineer, an economics expert and a bamboo processing expert. The expert team visited the primary processing factory of the company first, and the manager gave an introduction to the production process and stated the major technical problems.

Next, the expert team visited a composite board factory belonging to the company and had a detailed discussion with the local managing staff and technicians on each of the processing steps. After visiting the factory, the expert team produced a report that provided some suggestions to improve the production efficiency of the company.

A2.2   Problems and Solutions

The report that follows consists of the experts' suggestions for The Bamboo Company in Vietnam

and introductions to bamboo processing experiences in China. As a result of the communication between the experts and the company's management staff, the report provided suggestions on three aspects of the company that needed attention: the need for a sustainable supply of raw materials; the bamboo composite board processing technologies in use; and improvement of the company's production management. As details of processing techniques were discussed during the visit and the consultancy, the report does not repeat the details of each technology.

The first two of these aspects are discussed below; Section A2.3 discusses technical problems in processing in more detail; and Section A2.4 provides suggestions to the company's production management.

A2.2.1   Establishing a stable and sustainable mechanism for the supply of raw materials

The supply-chain model: 'company + farmer'

The Bamboo Company has played an important role in promoting the industrialization of bamboo in Vietnam by establishing bamboo processing factories. The development of the local bamboo sector will benefit the rural society, as bamboos are usually distributed in areas of poverty. Bamboo industrialization will consume a large amount of the local bamboo resources but, at the same time, it will add value to these resources and improve community incomes. If the local communities are aware of the significance of bamboo utilization, they will pay more attention to bamboo forest management and the development of bamboo plantations, which is also helpful for the improvement of the ecological environment. In conclusion, the company has made a great contribution to the development of the bamboo industry in Vietnam, as well as to the alleviation of the poverty of the local people in this poor mountain area and to the development of community economy.

However, with the development of bamboo processing in the past 1.5 years, and the lack of the company's cooperation with other bamboo enterprises, such as chopstick factories, the price of raw bamboo materials in Vietnam has risen significantly. Meanwhile, the company is facing the serious problem of instability and lack of bamboo raw materials. Although this is good news for local bamboo farmers, who may be encouraged to develop more plantations, it will have negative effects on the company's production, productivity and profitability. So it is very important for the company to investigate how it can establish a mechanism for a long-term and stable raw material supply chain.

One of the key methods of stabilizing raw material supply is to establish a working mechanism between the company and the bamboo farmers. Unfortunately, the experts learnt that most of the bamboo forests were not managed in a correct and scientific way. Some bamboo farmers managed bamboo in an unsustainable way by cutting down the bigger bamboos and keeping the smaller ones, mainly for short-term benefit, while some bamboo forests have been overcut and, in some cases, the young bamboos have been cut and the old ones left. If the farmers were trained to manage the bamboo forests, their yield and quality would be improved markedly and the farmers' income would increase too. In that case, the 1–2 year old bamboo would no longer be harvested and it would be unnecessary for the company to worry about the purchase of bamboo material older than 3 years of age.

In China, the situation is different. There is a partnership between processors and bamboo farmers, and the processors provide technical training on high-yielding bamboo cultivation. The two parties sign raw material supply contracts within which, when the farmers are establishing a new bamboo plantation, the company is responsible for providing technical guidance, and even free high-quality plantlets, so that the new plantation becomes a new raw material supply base for the company. Generally, the Chinese bamboo processing companies organize the contracted farmers to hold a meeting once a year, during which the farmers exchange their experiences on bamboo cultivation, and the farmer who cultivates the best bamboo forest is rewarded; the rewards can be cash, fertilizer and other honours. These activities should be done in cooperation with the local government, especially the forestry department. I think the local government likes to support the partnership, because it is beneficial for the farmers, which is a double benefit for the company and the farmers, and also helpful for economic development in a local area.

A2.2.2   Making the best use of the bamboo resource in the local area

Because the company only produces floorings, just the middle part and the middle lower part of bamboo poles can be utilized, and the other parts are all wasted. However, there are many bamboo chopstick and toothpick factories around, and these do not have to use the middle part of the bamboo, the other parts will do. Therefore, we suggested the following strategy.

The company could cooperate with other enterprises to improve the utilization rate of the bamboo culms, by different enterprises utilizing different parts of the bamboo culms. The company should aim at whole culm utilization, and make efforts to realize this gradually. This will help to reduce costs and increase the added value to the raw materials. Currently, both The Bamboo Company and the chopstick factories are facing the challenge of lack of resources, increased costs and less profitability. So The Bamboo Company should select one or two chopstick factories as partners to trial on whole culm utilization. If successful, the scheme could be spread. In order to extend this cooperation over a large area, it would be feasible to divide the existing primary processing factory into three parts. This would be convenient for providing material for the other chopstick factories, and also for reducing the transportation distance of the material and the costs. Alternatively, some suppliers could be selected to contract with the preprocessing factory, and then the company could focus on the quality control of bamboo strips, raw treatment and drying, so as to reduce the management costs. This would be a multiple benefit mechanism as it would be not only for the company itself, but also for other processors and suppliers.

So, we have two options for The Bamboo Company: (i) to extend its primary processing business by dividing the existing factory into three; and (ii) to contract the primary processing factory to the suppliers, with The Bamboo Company taking charge of the treatment and drying processes of the bamboo strips.

A2.2.3 Strengthening technical training and cooperation with the local government

During the investigation, the expert group found that nearly half of the bamboo preprocessing equipment did not get proper maintenance and was broken down. The efficiency of the workers was also quite low, not even matching that of a quarter of a Chinese worker. The experts also found that the company lacks a close partnership relationship with the local government. So the expert group then suggested that training be carried out for the company manager, technicians (employees) and local government officials.

Training for the manager

At least three managers should take part in the course 'International Training on Bamboo Industrialization Processing Technology' in China in September 2006, which is sponsored by the Ministry of Science and Technology of China (MOST), INBAR (the International Network for Bamboo and Rattan, now the International Bamboo and Rattan Organization) and CAF (the Chinese Academy of Forestry).

Training for the local officials

At least one and preferably two local officials should join the training in China in the current and/or following year. The Bamboo Company should recommend one official to take part in the training in September 2006; in that case, the Chinese government will cover the local expenses in China, if the local government or The Bamboo Company pays for the international travel.

The training for the local officials and technicians is very important. For this, Chinese experts could also be invited to Vietnam. The above-mentioned training activities should be done in cooperation with the local government and international organizations, from which The Bamboo Company could try to get more support, including financial support.

Training for the employees of The Bamboo Company

The purpose of this training is to enable the employees recognize the importance of working procedures. To train the technical staff in key procedures, a Chinese consultant will need to work with the technicians of the company and resolve the problems. The technicians will be trained during the course. A long-term programme on a consultancy service contract is advised as the consultant may need to work with the company for a long time to resolve the problems comprehensively. The other side of this is that the company would then be able to provide technical consultancies for the other factories in Vietnam.

A2.2.4 Adjusting the product structure

Three-layer composite flooring, which has good prospects in the international market is the only product of The Bamboo Company, but this single product structure has many risks and problems. Besides the market risk, the single product structure reduces the utilization rate of the bamboo raw material.

The current utilization rate of bamboo material by The Bamboo Company is 5–6% and this could be increased to 8% if the processing was to be improved. The utilization rate could be increased over 10% if the production was expanded to produce vertically pressed composite flooring and other products, and this would also be good for the profitability of the factory. The utilization rate could reach 40% if bamboo carpets and bamboo charcoal were produced as well, and this could also reduce negative influences on the environment. Further, if the parts of the bamboos not used by the factory are sold to the chopstick factory, the utilization rate will rise more to nearly 50%. Briefly then, the needs are to adjust the product structure, improve the utilization rate of the bamboo material by selling the so-called 'waste materials' to other factories, produce other products, reduce the costs to improve the profitability, and reduce the negative influences on the environment.

Compared with the bamboo products in China, the products of The Bamboo Company have a greater competitive advantage as the price of bamboo raw materials in Vietnam is only one third of that in China. Now therefore, besides improving its key processing techniques and equipment, the company should pay more attention to the improvement of enterprise management, adjustment of product structure, improvement of the production flow, and strengthening the cooperation with the farmers and local government.

A2.3 Technical Problems in the Processing of Bamboo Flooring and Possible Solutions

Composite bamboo flooring has been emerging in recent years, but there have been no effective solutions for the cracks and deformation of the flooring products that can occur to date. The Bamboo Company is more advanced in using *Dendrocalamus barbatus*, a sympodial bamboo, as the surfacing board for bamboo flooring. The expert group understands that using sympodial bamboos as raw material for flooring production is more difficult than using the monopodial bamboos. What is more, the equipment and blades that Chinese factories have used are mainly for *Phyllostachys heterocycla* var. *pubescens* (Chinese Moso bamboo), which is a monopodial species. So the exploration that The Bamboo Company has made on processing sympodial bamboo has played a positive role for world tropical bamboo utilization and application.

Table A2.1 lists some of the problems that the expert group has seen and discussed during their visit in Vietnam. All reasons given for the problems experienced by The Bamboo Company and for their solutions are given according to

Chinese experience in processing *Ph. heterocycla* var. *pubescens*, and they may not be suitable for sympodial bamboos. So it is necessary to carry out more exploration of the best processing techniques for the company. Here we have only concentrated on material feeding, gluing and sanding during the bamboo flooring production process. Some of the processing equipment mentioned is mainly used in a few factories with a large production capacity in China.

Some of the problems listed in Table A2.1 have been discussed during the expert visits. The following are some further comments for reference by The Bamboo Company.

Bamboo processing machines have lagged behind in their production and development, and neither the design nor the manufacturing techniques for the machinery are well advanced. At the same time though, the existing machines are quite selective in their ability to process different bamboo species, and so they are not universally applicable to all species. Consequently, engineers should explore their own methods for effective machine debugging according to the real conditions and their working experience.

The processing techniques used for the bamboo strips are most important in the whole process of bamboo flooring production. In the past, because strip processing is part of primary processing, people did not pay much attention to the quality and standardization of the techniques used at this stage. In recent years, many bamboo processing factories have begun to pay more attention to the quality of the strips produced, and the important step in strip processing is the 'fine planing'. Also, many problems in strip processing have been found to be directly related to the quality of strips, such as chinks of glue on the finished product, the strength of gluing, the size of the preprocessed panel and cracking at the edge of panels, etc. As a result, many Chinese bamboo flooring factories have now restudied the technology of strip processing.

A2.4 Suggestions to the Company's Production Management

A2.4.1 Present status of bamboo flooring in the world markets

After several days' visiting and investigating The Bamboo Company, the expert group thinks that

Table A2.1. Problems during bamboo processing by The Bamboo Company and suggested resolutions/causes.

No.	Item	Problems	Resolutions/Causes
1.	Bamboo selection	Strip shrinking Edge swelling and distortion after drying Colours not unified Strip cracking Damage by pests	Use bamboo culm between 3 and 5 years old
2.	Sawing	Breaking and cracking of the strips after sawing	Recommendations: fast feeding and discharging, and making sure that the sawn strips are immediately processed further and do not pile up in the open space for a long period of time Reasons: the saw is not sharp enough. low rotation speed of saw. improper operation of saw
3.	Cutting	Bamboo strips are too wide	Cut the raw bamboo as precisely as possible so saving bamboo raw material, reducing the cracking of strips and ensuring the quality of the bamboo surface
		Low utilization rate	Classify the bamboo culms by diameter and thickness Use a suitable vehicle to collect the raw materials and bamboo culms
4.	Planing	Cracking of bamboo strips Cracking on both ends of the strips (difficulties in feeding into the primary planer) Damage on the surface of the strips	Bamboo strips are too wide Too much moisture content in the strips Top part of the bamboo should not be used as the walls are too thin Strips are not straight enough Low rotation speed of the blades (increase the number of blades to 5, increase the blade rotation speed to 4000 rotations/min) The bouncing range of the feeding wheels is too small The spring of the press wheel is too stiff The tray under the blades is not properly positioned The cutting range is too large The blades are not sharp enough The strip stops during feeding The front angle of the blades should not deviate by <5°
5.	Steaming Drying Carbonization	The time between harvesting to processing is too long, as is the storage period in the factory	Drying must be carried out before the materials are attacked by fungi, raw materials should be put into the drying room within 60 h of harvesting Dry the material at a low temperature and in a strong air current Try to use twice-carbonization technology to avoid uneven colouring This technology should be applied at Lanshan Chemical Plant for three reasons: (i) to save drying time; (ii) to avoid a lot of storage; and (iii) to reduce transportation costs and increase product quality

Continued

Table A2.1. Continued.

No.	Item	Problems	Resolutions/Causes
6.	Fine planing	Bamboo strips are cracking	Bamboo strips have already cracked before fine planing
		Strip edges are partially broken	Bamboo strips are seriously distorted after drying
		Many spots on the strip surface	Front angle of the blades is not properly installed
		No right angles	High pressure at the feeding part
		Not smooth on front and tail parts	Pressing board or localizer is too tight
		Uneven sizes	Blades are not sharp enough; Structural problem with the equipment Cutting size is too big or too small, recommended size is 3 mm
7.	Grading and colour matching		Set a special process for grading and colour matching between fine planing and assembling
8.	Gluing and assembling	Not smooth on the surface of panels	Apply hot presser
		The thickness and width of the strips are not uniform	The equipment need to be adjusted from time to time to ensure the accuracy of the sizes of the strips
		The glue is not solidified	The power of the presser is not enough
		Fissures or shallow glue	Uneven gluing Time for glue solidification is not sufficient
9.	Sanding	The uneven surface of the bamboo strips affects the quality of gluing	A key to the solution is to increase the times of sanding – to sand twice before pressing, and sand a third time after pressing
		Variation in the thickness and width of the strips will cause cracking in the panel after pressing	
		The surface of the flooring panels is reduced too much	
		There is some sponginess or bubbles on the surface of the flooring	
10.	Other issues	There are some glue chinks emerging on the finished products	Too much glue used during the assembling or uneven gluing; Not enough time allowed for solidification
		Cracks after grooving	Cracking during the earlier processing procedures Blades are not sharp Variation in the thickness and width of the strips during pressing Overcutting Distance between the blades and tray is too large
		Soft surface of finished products	Reduce overcutting and over sanding for the green part of the bamboo, increase cutting and sanding of the yellow part Select high-quality lacquers with high rigidity
		Spots on the surface of finished products	Lacquering workshop should be separated from the other workshops Add a dust cleaner for the surface of the panels
		Surface of finished products is not smooth	Reduce the degree of sanding

Continued

Table A2.1. Continued.

No.	Item	Problems	Resolutions/Causes
		Drying problems: existing drying equipment is not suitable for bamboo materials	Recommended that orbital or successive drying facilities, which track the product slowly and automatically forwards through a very long drying furnace, and are of low cost and high efficiency, are used

it is a great company in the bamboo floor industry. The group thinks that it is absolutely right to choose 'composite bamboo floor' as the dominant product, as this totally fits the demand of the international market and avoids competition conflicts.

However, the expert group finds that there are still some issues that have to be improved for the company in overall management, mainly from two aspects: prices and products.

Prices

Bamboo floor products appeared in the world markets in the late 1980s. At the beginning, the international price was US$30/m^2 for 15 mm solid bamboo flooring, and the raw material price at that time was US$20/ton for Class A bamboo (in Chinese terms, there are 10 culms/ 100 kg). When bamboo floors became more accepted on the international market though, because of the lack of order in competition among enterprises, the price dropped dramatically and management patterns in the industries were improved.

At present, the FOB (Free On Board) prices for Europe and North America are as low as US$15/m^2, and even US$13/m^2, and the quality standards are higher. Moreover, in those main export countries for bamboo floor (e.g. China), the price for material is above US$90/ton, which is five times greater than before.

According to statistical data from Chinese customs, the amount of bamboo floor exported grows 30% yearly, while the value of export delivery stays stable. This shows that the market price is decreasing, and the competition is becoming fiercer.

Products

On analysing the product series and categories of flooring, we find that three-layer natural horizontally pressed bamboo flooring was most popular several years ago, while at present carbonized vertically pressed bamboo flooring is in greatest demand, followed by bamboo-timber composite flooring. In other words, looking into the future of the international bamboo floor market, the winner must have three qualifications: (i) have good quality and stability; (ii) be relatively low cost and with a low product price; and (iii) have high productivity.

A2.4.2 Suggestions for strengthening the company's production management

To obtain the product qualifications listed above, The Bamboo Company should be a leader in management. Management covers supervision, planning and product designing. First, the product series and specifications should be decided according to the market demand and material supply. Secondly, the company should choose the right site for the factory. Thirdly, the factory management model should be gradually improved. Lastly, the company should pick reasonable processing technology and equipment.

A2.4.3 Technology aspects

The technological problems that the company has to face in the short term are the following:

1. Controlling the age of its raw bamboo material.
2. Making full use of the raw material.
3. Improving the product quality rating in the processing, including how to increase the quality rating of semi-manufactured bamboo strips, how to strengthen their resistance to deformation, slipping, expansion and abrasion, and how to improve the rigidity of the lacquer used.

4. From the corporation management aspect, we think that it is necessary to adjust the management thinking, as the company has developed on such a large scale.

A2.4.4 Improve the product structure and increase the number of categories

We suggest the introduction of at least two categories:

1. Vertically pressed composite bamboo flooring.
2. Pressed bamboo flooring.

The purpose of producing these two kinds of flooring is to increase the utilization rate of bamboo material, as together they will use bamboo strips with relatively low quality. Moreover, in the future, the dominant products are likely to be vertically pressed composite structures, as according to our knowledge of market demand, vertically structured products are more popular than those with a horizontal structure. The prerequisite is the market!

Bamboo strips of 7–10 mm thickness are used for vertical pressing, and the middle lower part of bamboo culm can be processed into bamboo strips of 10 mm thick; 14 of these strips can be processed into two horizontally pressed boards or 4–5 vertically pressed boards (with one layer for the surface).

The above arrangements could increase the bamboo utilization rate by 100%.

A2.4.5 The salary system

We suggest that the salary system is adjusted to adopt a piece rate instead of a daily rate. More than 90% of employees will be satisfied with a piece rate salary, although several weak employees may object to it. Quality management should be especially strengthened by the piece rate salary system. Both quality and quantity examinations will be implemented, and there should be no top limitation for the salary. In this way, employees will improve their working efficiency, and the company will save costs.

A2.4.6 Waste material disposal

To deal with the waste from the factory, we suggest the addition of drying, treatment and polishing technologies in the preprocessing factory, and introducing vertically pressed composite products, which will reduce the waste. However, the total amount of waste is still too large. These wastes can be sorted into different categories: some can be sold to paper mills, some can be burnt as energy for the home factory, while most can be pressed as sawdust bricks and made into charcoal. The bamboo roots can be burnt directly into high-quality charcoal, and the top part of culms can be sold to a bamboo slicing factory or used for construction.

A2.5 Possible Future Chinese Cooperation with the Company

One consultation on company processing techniques and production management may not solve all of its problems, and new problems may arise while the company is making efforts to follow the above suggestions. Therefore, the expert team think that it is necessary to continue the cooperation between the Chinese experts and The Bamboo Company in the future.

First, the company should take urgent and powerful measures to establish consultancy and technology transfer projects, and invite experienced consultants from China to provide solutions to the existing challenges and problems in its processing technologies and production management. This will be a key step leading to the company's future success.

The consultancy can be carried out in two stages: (i) dealing with the present technical problems in processing, for example, the low quality of strips, the cracks and rough surfaces of the flooring boards; and (ii) comprehensive resolutions for the management of company, which include identifying a series of measures to ensure the long-term sustainable development of company. These measures will include the following:

1. The layout of the primary processing factory and the main processing factory should be redesigned, while a drying room and other associated facilities should be provided.

2. Product categories should be increased and processing techniques improved, including the installation of the necessary equipment for new products and the application of a standardized quality control system.

3. Plans should be made for gross production and unit production, and salaries paid to workers according to their productivities.

4. Comprehensive utilization of the waste materials should be undertaken.

5. A sustainable supply channel of raw materials should be established, and awareness raising and training for local government officials and technicians implemented.

The above consultancy activities should be carried out at different stages according to the urgency of need over an estimated period of 12 months. So it is necessary for The Bamboo Company and the Chinese experts to now negotiate the future mode of cooperation and identify a working plan.

Index